Leaching of Low and Medium Level Waste Packages Under Disposal Conditions

Synthesis of an international workshop organized by the
Commission of the European Communities (DG XII.D1) and the
Commissariat à l'Energie Atomique (CEA-IRDI-DERDCA), held
at the CEN-Cadarche, France, 13-15 November 1984

Leaching of Low and Medium Level Waste Packages Under Disposal Conditions

Edited by

M Dozol (CEA)
W Krischer (CEC)
P Pottier (CEA)
R Simon (CEC)

Published by GRAHAM & TROTMAN Ltd
for the Commission of the European Communities

Published in 1985 by Graham & Trotman Ltd,
Sterling House, 66 Wilton Road, London SW1V 1DE, UK

Publication arrangements:
L. EISEN, Commission of the European Communities
Directorate-General Information Market and Innovation

Softcover reprint of the hardcover 1st edition 1985

EUR 10 220

ISBN-13: 978-0-86010-814-6   e-ISBN-13: 978-94-009-4974-4
DOI:10.1007/ 978-94-009-4974-4

FOREWORD

Leaching tests are conducted according to different standards, procedures or recommendations, generally under conditions different from those of the disposal media. Such tests are generally useful for comparing various containment matrices to ensure maximum isolation of radionuclides, and allow waste conditioning processes to be optimized.

However, when defining release hypotheses to be taken into consideration for risk analysis, these methods can often lead to hypothesis which are either too optimistic or too pessimistic. The uncertainties basically involve the processes by which containers deteriorate during transport, handling and utlimate storage. Moreover, the interpretation of test results must be more clearly correlated with the known physico-chemical mechanisms involved.

In attempting to find an answer to this question, tests have already been undertaken, including the use of lysimeter devices or the recovery of waste materials after several years of storage. The aim of the seminar was to exchange views among experts in characterization, waste management and safety analysis for the purpose of defining the principal actions that could be undertaken or of reorienting ongoing studies in order to limit the uncertainties involved.

C O N T E N T S

INTRODUCTION

The following report presents the summary of a meeting of experts, organized jointly by the French "Commissariat à l'Energie Atomique" (CEA) and Commission of the European Communities (CEC) and hosted by the CEA. The general objective of the workshop was to gain a better understanding of the source term to be applied in the assessment of the long-term performance of low and medium active waste packages in an aqueous environment.

The transfer of radioactive nuclides and other substances from a solid waste form into the water is often generally termed "leaching". Most of the tests to measure leaching were devised with the aim to develop and optimize leach resistant waste forms. The IAEA and ISO leach tests will measure, to what extent a matrix is effective in preventing the release of waste materials but the resulting data have almost no bearing on the rate of release under repository conditions. More recently introduced leach tests, generally with static or near-static leachant and fairly high sample surface area/leachate volume ratios will, at least in strongly simplified manner, reproduce the effects anticipated in waste repositories.

The scope of the workshop was limited to low and medium active waste. There is no universally applicable specification for these two general waste categories, but in Europe it is mainly understood, that they contain all radioactive wastes except heat-generating and high level waste, spent fuel and, at the lower active range, mining and milling residues. The discussions therefore included wastes containing transuranic nuclides, e.g. solidified reprocessing sludges and concentrates.
In addition to the large number of waste types in the low and medium active categories, there are the different matrix materials and the variety in the shapes and materials of the outer containers to consider in any realistic description of the waste. The disposal routes taken into consideration here for these wastes are the various forms of shallow land burial and the disposal in mined cavities or geologic repositories. In the future, waste containing significant amounts of transuranics will be assigned to mined caverns or geologic repositories, whereas general low-level waste containing less than a few nCi/g of alpha-activity may be routed to shallow land burial, as in France, the United Kingdom and the US.

The overall philosophy governing all disposal concepts treated in this context is that of long term isolation by a multi-barrier system. The system and most of its components are required to act as effective barriers for such extremely long periods of time, that genuine performance testing is impossible. The proof, or at least a reasonable assurance of the safety of the multi-barrier system must be provided by risk or performance analysis. To assess the effectiveness of the system, the risk analysis will investigate scenarios, in which water or brine will reach the waste and initiate the release of radionuclides at various times after emplacement of the waste. The composition of the water in contact with the waste and in particular, the concentration and physico-chemical form of the released nuclides characterise the source term. The description of the source term is the starting point of the calculations to determine the effectiveness of the barrier system. The source term will depend upon the scenario defined for a certain period of time and therefore vary strongly as both natural and engineered barriers either slowly degrade due to natural evolution processes or rapidly fail due to disrupture events or human intervention.

A reliable and accurate description or model of the interaction between water/brine, the host rock, the various engineered barriers and the waste is a very difficult task. A large number of physico-chemical reactions have to be identified and correlated not only for the state of thermodynamic equilibrium, but also during transient stages. In most cases, the complexity can be reduced by making conservative assumptions which simplify calculations or reduce the number of variables. These assumptions can be made after investigating and testing the relevant interactions in appropriate experiments. Once the model is assembled, certain features or the entire model can be tested against experimental results, although here again the relatively short time during which it is practicable to perform tests, will severely limit the validity of such verification tests. Certain experimental methods of accelerating tests are acceptable, others doubtful. Moreover, modellists and experimentalists must take into acount, that any simple model and any individual experiment can only produce results representing one "typical" case in a distribution band.

A reliable prediction of the performance of waste repository barrier systems in general or individual waste package types in particular requires close cooperation between those, who have to model the alteration phenomena and those who have to measure and observe these effects in experiments. In bringing together scientists working in these two fields in this workshop, we primarily intended to identify the data required for a realistic description of the source term and to understand and agree upon the nature of experiments required to verify the results obtained from associated models.

SESSION 1

DETERMINATION AND EVALUATION OF DATA REQUIRED FOR RISK ASSESSMENT AND ANALYSIS

SYNTHESIS OF SESSION 1

Chairman: J.C. Alder, NAGRA Baden

Subject: Determination and evaluation of data required for risk assess-
            ment and analysis.

Objective: Identification of safety-related properties and mechanisms

Scope: Materials and performance requirements for waste packages
            related to generic repository siting conditions.

SUMMARY OF DISCUSSION

1.1. General philosophy of assigning disposal concepts to waste cate gories (as currently understood in Switzerland).

1.1.1 Many countries only have a limited choice of sites for radioactive
waste disposal but of different nature. The requirements for waste packages have to meet the general safety requirements applied to a site-
specific repository project.

In Switzerland it is proposed to allocate the waste packages for medium
and low level waste (excluding fuel hulls) to a rock-cavity repository
project in a pre-alpine geologic formation ($\sim$500 m depth). Crystalline
rock, unhydrite and marl are considered; the last has been studied in
detail in a recent disposal feasibility study (Project Guarantee 1985).
The option to dispose of the low level waste in a near surface repository
is still being kept open.

In Sweden a rock-cavity repository (SFR) has been studied by KBS. In
Germany, low and medium level waste, until 1979 was placed in the ASSE
salt mine (-800 m). In the near future the disaffected iron ore mine
KONRAD will be used for the disposal of radioactive wastes with insignificant heat production.

Shallow land burial of radioactive waste in trenches or mounds is generally more restrictive with respect to long-lived nuclide inventory.
Surface disposal facilities of various designs have been operating in the
US, France and the UK. Consequently, waste package data required for
safety analyses and later waste package specifications required for
disposal acceptance could be different from one repository concept or
project to another.

1.1.2 Barrier functions

The release of the radionuclides is normally limited by the use of one or
more barriers between the radionuclides and the environment, such as waste
matrix, waste packaging, disposal structure, backfill material and the
surrounding geologic medium. When the safety of a disposal facility is
evaluated, all the barriers have to be taken into account as a system and
potential positive and negative synergistic effects of the combination of
barriers have to be analysed.

A barrier may have a number of important functions such as:

- physical (hydraulic) barrier
- chemical barrier
- mechanical barrier
- thermal barrier, etc.

The most important application of the physical barrier is to limit or hinder the water flow through the barriers by using materials that have a low or zero permeability to water. An example of low permeability material is a clay buffer as the one being used in the Swedish SFR facility which guarantees that the main transport mechanism for the radionuclides inside the clay barrier is diffusion and not convection. Zero permeability is obtained by a closed metal container.

The chemical barrier effect depends upon solubility limitations in the chemical environment created by the barrier or upon ion-exchange and sorption effects in the barrier.

1.1.3 Repository design, safety system and mechanisms

A conceptual drawing of the Swiss "Type B" repository project is shown in Fig. 1.1., a view of a MLW disposal gallery in Fig. 1.2., the disposal units are containers of 20 m^3 volume grouping some 42 standard 200 litre waste drums (Fig. 1.3.) embedded in cement.

These monolithic rectangular units are closely packed in the repository and all free spaces between and around them will be backfilled with a plasticised cement grout.

As the repository itself is concrete-lined, the engineered retention system will consist of five diffusion barriers (Fig. 1.4) and the impermeable, but ultimately corroded steel drum. The designed features listed were chosen for the feasibility Guarantee Study ; no optimization has yet been carried out. The design objective of the engineered barrier system is to assure, that under all circumstances, release of radionuclides is governed by diffusion mechanisms and not by convective or other flowing transport effects.

1.2. Scenarios and relevant waste features

1.2.1 Scenarios

In the safety analysis carried out by NAGRA in Switzerland the reference scenario (normal evolution) presumes that the near field environment of the waste is saturated with water shortly after closure of the repository. In this case convective flow through the repository is prevented by the backfill. Consequently the principal transport mechanism in the near field, is diffusion through the subsequent diffusion barriers (Fig. 1.4.) and no very strong leach requirements are placed upon the waste package.

In the Swedish safety study (and in WIPP - Ed.) swelling by water uptake and pressurization by radiolysis or other gas-generating phenomena may ultimately lead to the rupture of one or more engineered barriers.

Due to the favourable geology, the KONRAD iron ore mine, one of two repositories planned in Germany, will accommodate waste, with insignificant
heat production, e.g. from reactors, research establishments, industry and
the decommissioning of nuclear facilities. The KONRAD mine is practically
impervious to water and the principal concern of safety analysis is the
release and dispersion of activity by a fire or a mechanical impact (drop)
in the pre-closure stage.

The Gorleben facility, a deep geologic repository in a salt dome, will
accept all types of radioactive wastes including high level and alpha-
bearing wastes.

For this site, the scenario of a brine intrusion during the post-operational phase and subsequent expulsion by salt convergence has not been
precluded.

In the three mentioned safety assessments, the amount of water/brine in
contact with the waste will be small for the reference post-closure
scenario, so that the activity transported is governed by the solubility
of the radioactive species.

The French approach for shallow land burial is different. After an acci-
dent such as severe earthquake leading to rupture of the cover, the R.N.
movement around the waste could become less restricted and the release of
activity is then controlled by the leach rate.

Risk analysis will be based on release rates adjusted to specific site
conditions by correlating the leaching mechanisms identified in laboratory
experiments to describe the real phenomena in a disposal trench, taking
into account the water chemistry, the characteristics of the concrete
structures, the various other barriers and the time laws for container
degradation.

The respective laboratory leach rates can be related to the in-situ
release rate by a "tranfer function", which could, to a certain degree, be
verified by lysimetry tests.

For near surface disposal the scenarios resulting in the highest releases
and exposures are those involving intrusion/destructive human intervention
after the period of control (300 yrs). This also applies to the DRIGG LLW
disposal (UK) after control over land use.

1.2.2 Relevant characteristics and properties

Mr. Alder defined characteristics as the invariable attributes of the
waste package or the repository system, whereas the properties describe
certain parameters which can vary under different external conditions.

Thus, the package dimensions, matrix material, waste composition and
inventory will be considered as characteristics, whereas leachability,
mechanical strength, corrosion resistance will be described as properties.

To be able to benefit from a barrier effect in the safety assessment it is
essential to be able to predict all processes that could potentially
offset or diminish the barriers' function. Such processes could be
natural ageing processes or interaction from the other components of the
disposal system by mechanical or chemical action. The behaviour of the

waste package can in this context be of paramount interest. In the planning of testing and characterization of the waste packages it is therefore essential to make a model of the disposal safety concept to find out which data are most important. The best way is to refer to results of a corresponding safety analysis. Different cases for the relevance of the properties and characteristics are to be considered. In the following some examples are given.

Waste properties affected by other barriers

A waste package consists of the radioactive material, a matrix and a container.

The form of the radioactive material constitutes the first barrier in the disposal system. Three examples with different properties could be mentioned:

a) ion-exchange resins where the activity is weakly bound to the resins and where the composition of the water getting into contact with the radionuclides determines the rate of release; for instance in concrete water cesium is released very quickly;

b) coprecipitation sludges, where the radionuclides have been rendered almost insoluble by pretreatment; also in this case the water composition is very important; the release rates are generally lower;

c) metal with induced activity, e.g. core components, where the corrosion of the metal and the solubility of the radionuclides determine the release rate; an example of very low release rates is given by nickel from stainless steel in concrete.

The conclusion is to keep in mind that the source-term in leach experiments of interest is the release rate of the radioactive material in the chemical environment created by the other barriers.

The matrix is commonly described as a means to immobilize the radioactive material. This is either achieved by a pure encapsulation of the radioactive particles, as is the case for incorporation into bitumen or polymers, or by combination of encapsulation and chemical reaction as is sometimes the case for incorporation in concrete.

In the final disposal the encapsulation acts as a barrier to water penetration. The long-term function of this barrier is dependent on processes, such as swelling and gas generation or external chemical attack that could alter the material. For example, the uptake of water in a dehydrated salt or resin in bitumen could lead to swelling, which could break the bitumen film. Concrete could be leached by the ground water, which would change its porosity. Both these effects and the ageing process change the transporting capacity of the matrix.

The container is primarily intended to facilitate the handling of the waste package and to prevent release of radioactivity during handling operations. In the final disposal the container for low and medium level wastes has seldom been included in the safety analysis. However, in certain cases, such as a closed metal or a concrete container, the container will serve as a barrier for a certain period. The length of this period is determined by the chemical environment inside and outside the container.

Waste characteristics affecting other barriers

As has been shown above the behaviour of the waste package in the disposal is strongly related to the function of the other barriers and the water chemistry. The waste package could, however, also affect the other barriers of the disposal system either mechanically or chemically.

Mechanical effects could be caused by swelling or collapsing or by gas production, e.g. by the corrosion of metallic containers. This behaviour will primarily affect the physical function of the other barriers, that is the limiting or hindrance of water transport. Physical changes could also be the result of a chemical interaction, e.g. $Ca(HO)_2$ on clay.

Chemical effects could be created if the waste release substances that diminish the retention capacity of other barriers, either by changing the chemistry (e.g. release of $Ca(HO)_2$ from concrete would affect clay and rock retentional properties) or by release of complexing agents (e.g. from decontamination or laboratory effluents).

The release of a chemical species is also dependant on the transport of these species or its reactions in the next diffusion barrier. Example: when concrete is in contact with sodium-bentonite, calcium ions will be released and react with the sodium-bentonite. The bentonite thus creates a sink for the calcium ions which could increase the release rate.

1.2.3. Ranking of waste package characteristics /properties

A general order of priority/merit of waste package characteristics and properties covering all waste categories and repository concepts, is not possible.

Within the scope of this meeting, however, the implications for the leaching release mechanisms in low and medium level packages under disposal conditions can be identified as relatively important when water convection is assumed to take place in the repository.

The Tables I.1 and I.2 present an attempt to identify relevant properties of waste packages on the basis of the following considerations:

Normal evolution

In all repositories for (non heat generating) LLW and MLW the waste form will either remain dry (salt environment) or with slowly infiltrating water. Significant liquid flow rates and hence advective nuclide transport are prevented by the backfilling impermeable host rock. Diffusion transport of nuclides out of the repository will be so slow, that the predominantly short-lived soluble activity will decay before entering surface waters. Further retardation by sorptive mechanisms can be expected in most cases.

Since release is determined by diffusive transport, the decrease in integrity of the barriers, including the waste matrix, with time will lead to an enhanced radionuclide transport. This can be seen in Fig. 1.5 where the near-field release rates resulting from the modelling of Fig. 1.4 for the Swiss repository type B project in marl show marked increases

at 500 and 10 000 years at which times partial degradation of the
barriers is assumed.

In shallow land burial much of the protection afforded by the earth cover
and the concrete structures is assumed to become less effective after a
period of institutional control ranging from 100 to 300 years and could
eventually be destroyed completely, e.g. by excavation.

Disruptive events

The performance assessment of the barrier system must also take into
consideration conceivable events not anticipated as part of the normal
evolution.

Such occurrences include natural catastrophies such as severe earthquakes, flooding of the site and human interventions like drilling,
mining, or in the case of shallow land burial, aircraft crashes, during
the controlled period.

One of the consequences of these events can break the external barriers
so as to admit a significant water flow through the repository. The
release mechanisms will change from static dissolution to dynamic leaching, the diffusive nuclide transport will become advective.

The release of all radionuclides increases considerably, as can be seen
from Figs. 1.5 and 1.6, where both diffusive and convective transports
have been modelled for the Swiss repository type B design in marl.

The properties listed in the tables only concern the release of nuclides
by water transport in the post-closure phase of the disposal facility.

PROPERTIES OF LOW AND MEDIUM LEVEL WASTE PACKAGES RELEVANT
FOR RELEASE BY WATER TRANSPORT UNDER DISPOSAL CONDITIONS

Table I.1 NORMAL EVOLUTION

1.1. **Waste properties affecting the barrier system and disposal conditions**

1.1.1 for all disposal options:
- corrosivity, other potential chemical interactions
- absence of free liquid
- radiation effects on the near-field of the waste package

1.1.2 for geologic disposal in deep salt, clay and rock
- gas generation by radiolysis, chemical or biological mechanisms

1.1.3 for hard rock
- dimensional stability (swelling, shrinking, collapsing of packages)
- compressive strength
- reaction of waste or matrix materials with the backfill (e.g. cement and bentonite)

1.1.4 for shallow land disposal
- dimensional stability
- compressive strength

1.2. **Properties governing nuclide release from the waste forms**

1.2.1 for all disposal options
- corrosion resistance of matrix
- solubility of the nuclide species in repository conditions
- solubility of waste salt and matrix materials
- hydraulic conductivity, porosity, pore size distribution of solid waste form
- diffusion and sorption/desorption data

1.2.2 for shallow land burial
- leaching meachanisms for advective conditions, transient flow
- potential microbiological transformation of the waste

Table I.2 DISRUPTIVE EVENTS (EARTHQUAKES, HUMAN INTRUSION, ETC)

2.1.   Waste properties affecting the barrier system
       less relevant as barriers are breached from outside

2.2.   Properties governing nuclide release from the waste forms
       (first priority: low nuclide release)

2.2.1 for all disposal options
       - concentration of radionuclides and homogeneity of distri-
         bution
       - leachability (including dynamic conditions)
       - mechanical strength and hardness.

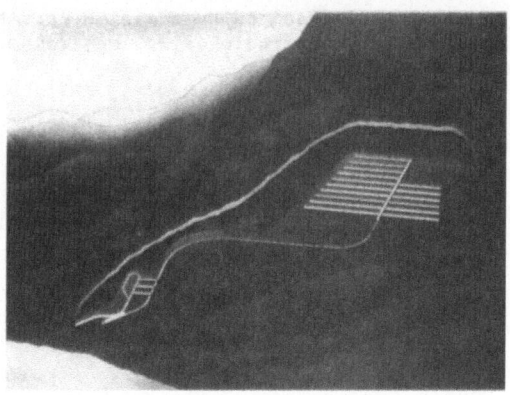

Fig. 1.1 SWISS "Type B" Repository Project

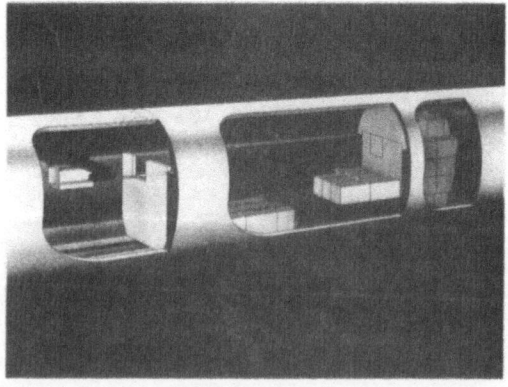

Fig. 1.2 View of a MLW Disposal Gallery

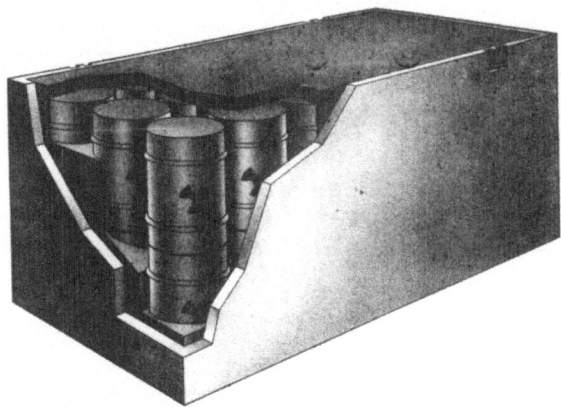

Fig. 1.3 View of a container (Disposal unit)

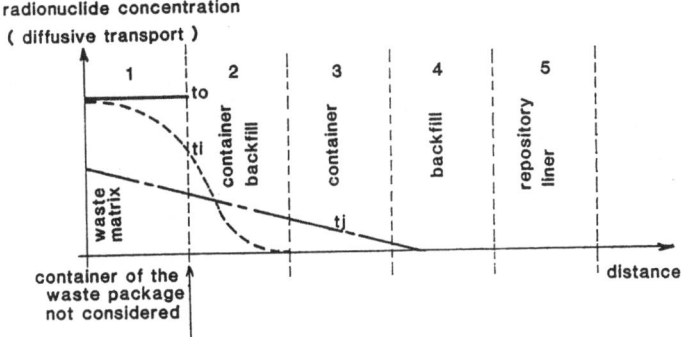

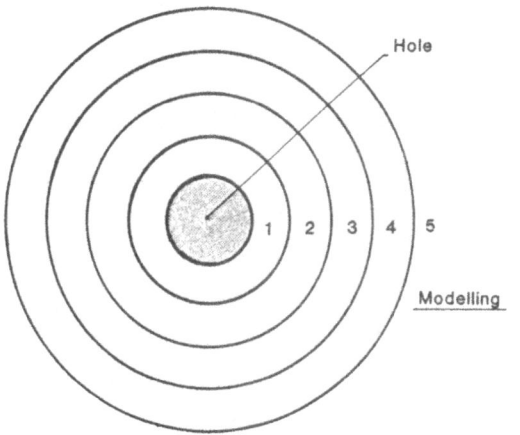

Fig. 1.4 Engineered Retention System

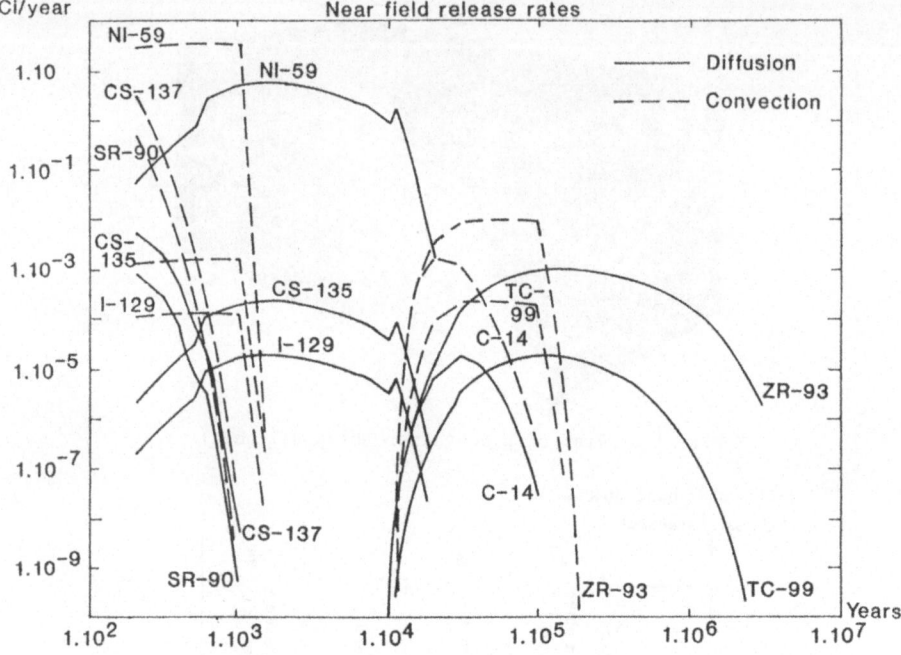

Fig. 1.5   Near Field Release Rates For Beta Gamma Nuclides

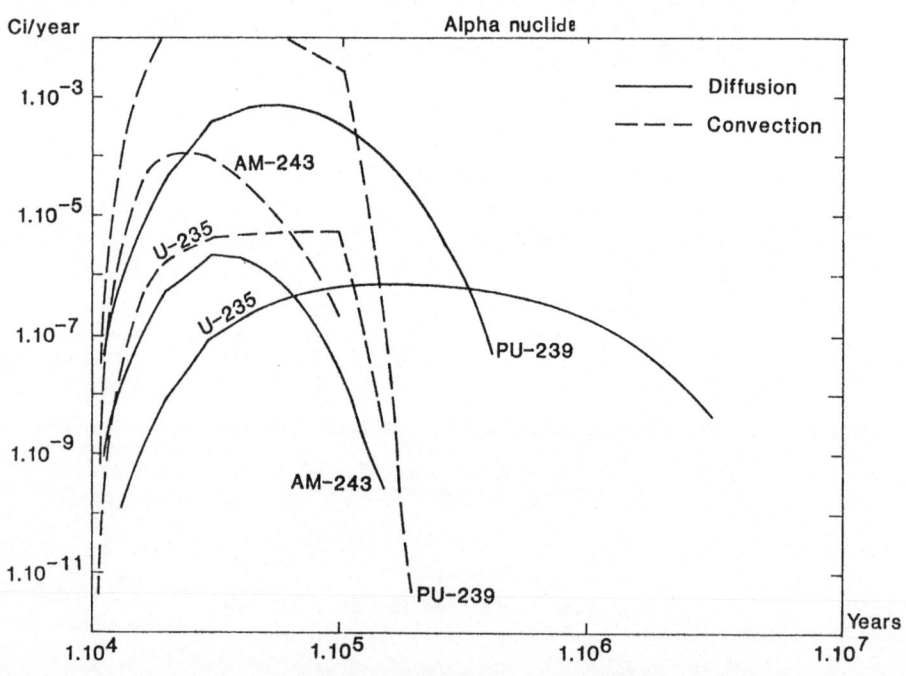

Fig. 1.6   Near Field Release For Alpha Emitters

SESSION 2

INTEGRATION OF WASTE PACKAGE CHARACTERISTICS AND
ENGINEERED BARRIER INTERACTIONS INTO CHARACTERIZATION
AND LONG-TERM BEHAVIOUR PROGRAMMES

SYNTHESIS OF SESSION 2

Chairman: Dr. K. Brodersen, NRL Risø

Subject of discussion: Integration of waste package characteristics and
                        engineered barrier interactions into
                        characterization and long-term behaviour
                        programmes.

Objective: Appraisal of the time dependent interactions between barrier
            materials for repository safety assessments.

Scope: Elaboration of the physical stability of the waste form, of
        the chemical interactions with the surrounding barriers and
        the experimental approach to study release mechanisms.

SUMMARY OF DISCUSSION

2.1. Importance of phenomena for activity release in multibarrier systems
      (Dr. K. Brodersen)

Some examples of barrier systems are shown schematically in Fig. 2.1. The
first interaction level is that between the waste and matrix. The figure
shows examples of external water compositions in various disposal formations, but the internal chemistry of the system may, for a long period, be
entirely dominated by leaching from the waste (soluble salts), concrete
(high pH) or corrosion of steel (Fe^{2+} under reducing conditions). The
potential for transport of dissolved activity through defect barriers
(cracked or corroded) would depend on whether or not hydraulically conducting channels exist in the barrier interfaces. The possibility of
using bitumen as a sealing material between waste units has been examined
at Risø. The aim was to improve retardation in the transport of leached
radionuclides out of the system.

The effective diffusion coefficients for transport in pores (not bulk-
material) may include:

- Physical adsorption on pore walls
- Chemical retardation effects (ion exchange, precipitation)
- Effects of tortuosity and pore size
- Effects of dispersion in complex pore systems
- Ion exclusion.

For transport mechanisms, porosity and diffusivity are of major importance. Table II.1 gives some typical values of these parameters for various
barrier materials. Starting from the inside of the waste package, the
properties of the waste itself may constitute a barrier against release,
e.g. insoluble sludge particles, activity in activated steel components,
etc.

The second barrier is the matrix material which in many cases is employed
to convert the waste particles into monolithic blocks. It is important to
remember that low- and medium-level conditioned wastes are, unlike to high
level waste glasses, heterogeneous in nature. They consist of waste
particles (salts, sludges, resins, etc.) dispersed in the matrix material.

Lack of compatibility between the waste particles and the matrix can, under certain circumstances, constitute a problem. For example, the soluble salts or dehydrated ion-exchange resins are of salient importance in determining the behaviour in contact with water. The waste loading is the determining parameter in this respect.

2.1.1 Irradiation effects

The destructive effects of irradiation were illustrated with the example of ion exchange resins (IER) in concrete. Irradiation to a gamma dose of 10^8 rad and subsequent immersion into water, leads to the complete collapse of the matrix structure. This was attributed to swelling and cracking due to radiolytic gas formation (H_2). It was observed that the pH of the leachant changed rapidly from 7 to 4. The same effect was observed in Sweden at doses of 10^7 rad and in the U.S. with IER immobilized in polystyrene. KFK measured a change of pH when irradiated polymers were immersed in water. It is assumed that this is due to the splitting off of sulfonic groups, leading to swelling and to trapped ions in the matrix, which react with the protons of the leachant.

However, it was pointed out that the phenomenon varies with different waste types, matrix materials and waste loadings, e.g. polymers show enhanced water transport during irradiation, the leach rate of Magnox waste in bitumen increases under irradiation, a cement matrix containing borates disintegrated at a dose of 10^8 rad, whereas nitrates are retained safely.

In France, experimentalists have found that the H_2-production and consequent swelling of IER solidified in bitumen is depending on the degree of saturation of the IER. For this very reason the waste managers in France decided not to immobilize IER in bitumen. Mineral salts seem to be less troublesome. The degradation of nitrates under irradiation leads to O_2-release. The G-values for beta/gamma and alpha-irradiation differ apparently only by a factor of 3 to 4 (higher for alpha-irradiation). If the irradiation is carried out in a nitrogen atmosphere the degradation seems to decrease. The controlling factor of enhanced leaching due to radiolysis gas formation for cement is the specific gas permeability (pore structure), the geometry and accumulated dose of the specimen.

2.1.2 Water uptake

The first tests with full-scale bitumen waste forms carried out in Cadarache indicate an upper limit of 3 to 4% of water uptake. For bituminized waste the initial water content seems to be decisive. The specific surface of the specimen is thought to be of secondary importance.

Up to now only unrestrained tests have been performed, but for full-scale effects the restrained case (with surrounding container) should be investigated to check the effect of pressure build-up, which could be due to:

- Formation of radiolysis gases
- Water uptake
- Corrosion or degradation gases (CO_2).

It was pointed out that the production of CO_2 from bitumen in a microbial enriched environment can be as high as 100 ml/g (at 32°C, aerobic conditions, in liquid phase and with addition of yeast extract).

Areas of particular interest with respect to water uptake are:
. Rate of water uptake
. Maximum swelling pressure
. Engineering countermeasures.

The flowing of bitumen or the cracking and crumbling of cement products
may also need to be evaluated with regard to the specific surface used in
source-term calculations.

2.2. Existing modelling capabilities

Water transport through the repository system, has been discussed in
Session 1 (para 1.1.2). For the modelling of release mechanisms and
extrapolations in time it is generally assumed that the waste form is
physically stable. Any reduction in its long-term stability has to be
compensated for by appropriate countermeasures, i.e. by reliability of
the outer barriers (container, overpack, backfill, etc.).

2.2.1 Waste container

For non-alpha-bearing wastes one can largely rely on the resistance of the
container, which should last (assuming bulk corrosion) for the period the
waste must be isolated from the biosphere. Mild steel in an appropriate
thickness, as actually considered in the U.K., appears to be more reliable
for long-term predictions and thus easier to model compared to alloys,
like stainless steel, since only bulk corrosion has to be considered.
Pitting corrosion, the most likely and aggressive type of corrosion of
alloys, is difficult to predict. Of particular importance for the estima-
tion of external corrosion rates is the pH evolution of the groundwater.

In the Federal Republic of Germany two types of container corrosion
experiments have been carried out under full-scale, in-situ conditions in
addition to laboratory tests:

a) Empty mild steel drums of 200 l volume (Fig. 2.2) were stored for 13
   years in the Asse salt mine, under the conditions indicated in Table
   II.2.

b) Representative bench-scale samples for sphero cast iron drums of 270 l
   volume (Fig. 2.3) were stored for intervals of 3, 6 and 12 months in
   the Asse salt and the Konrad iron mine, under the conditions indicated
   in Tables II.3 and II.4. Interaction of spent IER and PWR evaporator
   concentrates embedded in a cement matrix and contained in cast iron
   cannisters were tested (4).

Shallow pitting corrosion amounting to a depth of up to 15 µm/year for
salt-brine was observed in all cases (Figs. 2.4 and 2.5).

2.2.2 Cement structures

No cracks have so far been observed in concrete drums stored for 13 years
in a trench of the "La Manche" site. However, some cracks were already
present before emplacement, since these drums had been stored on the
surface previously. They contained wastes conditioned differently from
todays practice. For this reason and also because these drums originate

from one trench only, it is rather difficult to draw general conclusions. First cracks in concrete trenches of the "La Manche" site were observed approximately 15 years after the closure of the trenches. Experiments to study the water permeability of concrete and cement in contact with clay are being carried out with tritiated water in Fontenay-aux-Roses and Risø.

2.3. Chemical interactions

In order to determine activity releases and migration mechanisms the chemical interactions between the different barriers have to be studied thoroughly. It must be stressed that engineered barriers can provide a considerable degree of protection for each other. Examples are steel/concrete, where the steel may strengthen the concrete or prevent leaching, while the high pH derived from concrete, passivates the steel. Mutual protection can also be expected for bitumen/concrete if the high pH prevents microbial attack of the bitumen. Two different types of concrete in contact with each other, like concrete containing silica fume and Densit/ordinary concrete, may improve the possibility of self-healing of cracks.

However, negative influences or at least complications for long-term assessments may arise due to the rather complex development of the water chemistry in the system with time. This implies effects on Kd, speciation and solubility.

Clay acts as an ion exchange medium. This may be reason why for bitumen matrices the rate of attack has been observed to be enhanced when clay was added to the leaching water. In the case of cement matrices the curing time seems to affect the retention capacity. At Fontenay-aux-Roses the interstitial water was squeezed out of cement paste and analysed . Three different types of cement were investigated:

. Portland cement (OPC)
. Pozzolanic cement (PZC) and
. Blast furnace slag cement (BFS).

The pH of the interstitial water was determined to amount to > 11 (due to the alkalis released). This means that leachates in the presence of cement and under low flow conditions will obtain a high pH. The yield of the alkalis Na and K measured in the interstitial water is reproduced in Fig. 2.6. This is in agreement with the Saclay measurements for concrete (full-scale, static leach tests as described in Session 3, para 3.2.2.1), a pH of 11 to 12 was obtained. This depended on the waste contained, but never fell below pH8.

2.4. Experimental methods to study release mechanisms

The outer barriers (container, backfill, concrete structure, etc.) have a double function:

a) To prevent or retard the penetration of water from the outside to the waste form and

b) To retard the migration of leached radionuclides out of the system.

For safety assessments it is desirable to model these two phenomena reliably in space and time. Currently only approximations can be made because the time dependent interactions between the barrier materials are poorly known. The water chemistry in a system consists of complex functions of transport properties and reaction rates for the various materials involved. The possibility of transport of dissolved activity through defect barriers (cracked or corroded) may depend very much on whether hydraulic conducting pathways exist within the repository, e.g. at the interfaces between the barriers. In more easily calculable, ideal cases, transport can be regarded as purely diffusive.

In general, one should aim for a compromise between integral experiments, which are difficult to interprete and simplified experiments, which are less representative.

2.4.1 Cement based matrices

The description of the release mechanisms as a function of time (source term) is particularly difficult in the case of cement based matrices. In order to obtain the upper diffusion coefficient, KFK performed leaching tests with pre-corroded specimens (accelerated approach, granulated samples in Q-brine at 40°C, see also Session 3, para 3.2.4). The test results suggest that for the first 100 years, diffusion is the prevailing release mechanism. After that it can be assumed that due to the complete corrosion of the matrix the predominant release mechanism becomes controlled by solubility only. This theory will be examined for various radionuclides, in particular for actinides. It is admittedly a rather pessimistic approach since in reality any presumption of release assumes a corresponding intrusion. However, it avoids complex and time consuming assessment procedures.

Leaching of Co-tracered cement specimens in distilled and in ground water indicated higher leach rates for ground water. ALso it was observed that Co migrated readily through soil due to complex formation. In addition it was pointed out that container corrosion products can affect the leaching of specific isotopes. The relative importance of the different effects has still to be determined.

2.4.2 Polymer and bitumen based matrices

Casaccia carried out leach tests in deionized water with different wastes incorporated in polyester resins. The results indicate that the diffusion coefficient depends on the waste type and loading and can vary by 2 to 3 orders of magnitude (Table II.5). One should, however, be cautious with releases based on deionized water, since the exchange mechanisms are rather inefficient with a low ion potential.

For bituminized waste an extensive test programme has just been launched in France. First of all, the bitumen itself is being thoroughly characterized. This includes an investigation of the radiation resistance, the identification of micro-organisms which cause microbial attack (first results are expected in 1 to 2 years time) and the associated gas formations (H_2, CH_4, CO_2, etc.) and degradation mechanisms. In addition samples will be stored in soil for 10 years.
Mol examined two types of bituminized waste:

a) Eurobitumen containing reprocessing sludges (sodium nitrates) from the EUROCHEMIC plant,

b) Bitumen containing effluent treatment sludges from the research centre of Mol.

Waste form (a) was leached in distilled water and in a clay water mixture. In both cases swelling was very pronounced (sometimes up to 540%) and a rapid release of nitrate was observed. The swelling is, however, largely dependent on the dimensions of and constraints on the specimen. Waste form (b) was leached in synthetic sea water at 500 bar during one year. In this case no swelling at all was detected. The pressure effect on the release has not been determined yet.

Storage temperature is an important design criterion for bituminized waste destined for granite repositories. A maximum storage temperature not exceeding 20°C below the softening point of bitumen (for Mexphalt 90/40 = 140°C) must be assured. The geological temperature has to be taken into account (at 300 to 400m depth it is ~55°C). The leaching behaviour has to be tested at the respective operational temperature. In France no change was measured in the viscosity of bitumen exposed for one year to 50°C. Similar tests are now performed at 90°C. The softening temperature depends on the waste content.

2.4.3. Effects of chelating agents

Complexing agents can mobilize radionuclides, in particular Co. For this reason, regulations may impose a limit on chelating agents in bitumen matrices (is in France under consideration). The amount of chelating agents present is dependent on the waste producers' declarations.

The limit for disposal of chelating agents is 8 wt.% at the US-Barnwell, South Carolina disposal site. All wastes having a chelating agent content between 0.1 and 8 wt.% must be solidified and segregated from other wastes in US-disposal trenches. There is, however, no upper limit at the Hanford, Washington site, although wastes containing more than 1 wt.% must be segregated from other wastes in the disposal trench.

2.5. Conclusions

For long-term predictions and the modelling of release mechanisms it is essential that the waste form is physically stable for as long as possible. Any reduction in its long-term stability must be compensated for by countermeasures, e.g. by reliability of the backfill material. Water transport through the barrier system depends mainly on the diffusivity and, in the case of convection, on the extent of defects in the barrier materials involved.

Regarding the extrapolation of experimental data in time it is questionable whether there is much to be gained by developing more complex models. Experimentalists have to aim for a compromise between integral experiments, which are difficult to interpret and simplified experiments, which are less representative.

2.6. References

(1) W. Hauser, Optimierung der Aktivitätsbeladung von schwach- und mittelaktivem Abfall, KFK 3825, Januar 85.

Table II.1 : Typical Porosities and Diffusion Coefficients
for Solutions in Pores

	Porosity ε	Diffusion coefficients (cm^2/s)	
		Water	Cs^+
Steel	0	0	0
Bitumen	0.005	$\sim 10^{-8}$	$10^{-8} - 10^{-13}$
Concrete	0.1-0.01	$10^{-7} - 10^{-8}$	$10-8 - 10^{-10}$
Clay	0.3	$10^{-6} - 10^{-7}$	$10^{-7} - 10^{-9}$
Water	1	2×10^{-5}	2×10^{-5}

Table II.2 : Storage Conditions for the Drum

Storage in Asse Salt mine (-750 m level)
storage temperature: 32°C

- Storage in rock salt
  Drums were covered with crushed salt

- Storage in a tunnel
  Drums were free standing in the dry atmosphere

- Storage in salt brine
  Drums were stored in a brine pool
  with the following composition (g/l):

Ca^{2+}	0.013	F^-	0.07
Mg^{2+}	109.0	Cl^-	295.0
Na^+	3.5	SO_4^{2-}	42.3
K^+	3.5		

pH - value   : 5.5

Oxygen content   : 0.15 mg/1

Table II.3 : Test conditions

Material: Sphero cast iron GGG 40.3

Specimen size: 10 x 20 x 40 mm

Test periods: 3, 6, 12 months

Triplicate samples per test period

S/V-ratio: Test conditions for internal corrosion: 0,1 cm^{-1}
                                   In the container: 0,15 cm^{-1}

Analytical methods: Gravimetry
                                   Surface diagrams
                                   Optical surface measurements
                                   Metallography

Table II.4 : Internal corrosion

- Bead resins: $LiOH/H_3BO_3$ spent ion exchange resins
                    S100/H500, residual humidity ca. 50 %, RT & 50°C

- PWR-concentrate: Evaporated, residual humidity ca. 10 %, pH $\approx$ 6-7,
                    RT & 50°C

- Cement paste: Aqueous solutions of PZ35, pH $\approx$ 12, O_2- content
                    ca. 6 mg/l, RT, 50° & 90°C

Outer corrosion media

- Dry rock salt: Asse in-situ, water content ca. 0,1 %, 28°C

- Salt solution: Q-brine, pH $\approx$ 5-6, O_2-content ca. 3 mg/l,
                    50°C & 90°C

- Dry iron ore stowing: Konrad in-situ, ventilated section,
                        1250 m level, 32°C, relative air humidity 60 %

- Humid iron ore stowing: Konrad in-situ, closed section, 1250 m level,
                          42°C, relative air humidity 82 %

- Konrad deep ground water: Solution with 14 % NaCl, pH $\approx$ 6,5,
                            O_2-content 5-6 mg/l, 50° & 90°C

- Roofed free space: in-situ HDB

Table II.5 : Diffusion Coefficients of various Isotopes
from Polyester-Resins incorporating MLW

Isotope	Diffusion coeff. (cm^2/day)	Diffusion coeff. (cm^2/s)	Fitting error (%)	Incorporated Wastes
Co58	5.296 E-06	6.129 E-11	10.5	Borates from PWR
	1.549 E-06	1.794 E-11	9.2	Borates from PWR
	6.773 E-08	7.839 E-13	16.3	Sludges
	5.879 E-08	6.805 E-13	12.3	Sludges
	7.040 E-07	8.148 E-12	5.5	(A+B) EUREX
	4.773 E-06	5.525 E-11	4.2	(A+B) EUREX
	1.968 E-09	2.277 E-14	5.8	MLW TRISAIA
Cs137	4.723 E-09	5.466 E-14	28.4	Sudges
	7.320 E-10	8.472 E-15	22.1	Sludges
	6.617 E-06	7.658 E-11	8.4	(A+B)EUREX
	1.584 E-07	1.833 E-12	4.5	MLW TRISAIA
	6.845 E-05	7.922 E-10	13.0	Borates from PWR
	6.985 E-05	8.085 E-10	7.9	Borates form PWR
Sr85	1.044 E-09	1.208 E-14	13.3	Sludges
	2.868 E-09	3.320 E-11	11.0	(A+B) EUREX
	1.762 E-08	2.039 E-13	26.8	Borates from PWR
Eu152	1.204 E-09	1.394 E-14	5.3	MLW TRISAIA
Ru106	1.187 E-09	1.374 E-14	3.9	MLW TRISAIA
Sb125	9.079 E-10	1.051 E-14	6.4	MLW TRISAIA

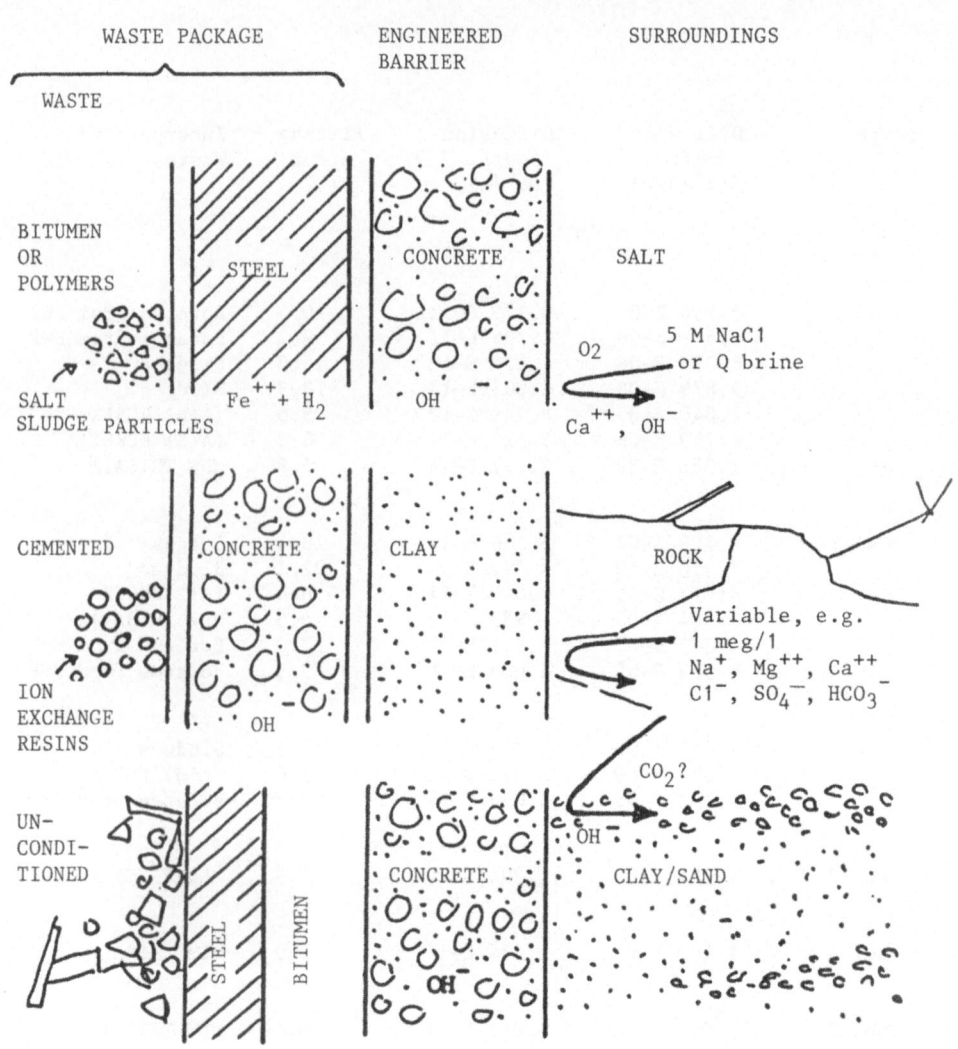

Fig. 2.1   Examples of Multi-barrier Systems

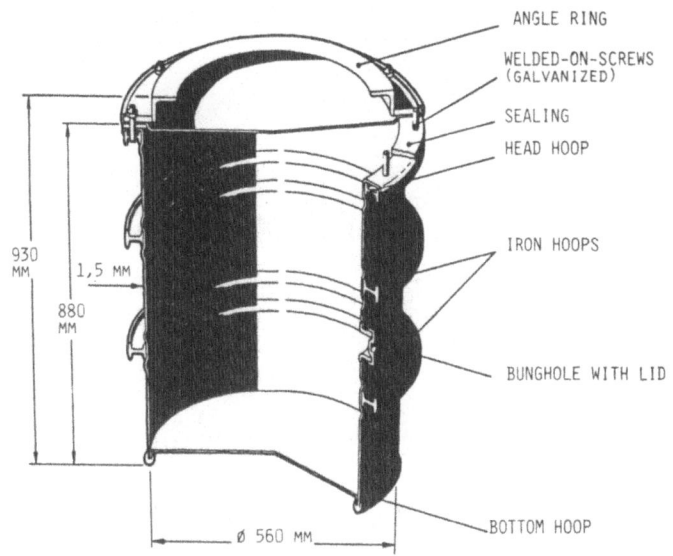

Fig. 2.2   200 Liter Iron-Hoop Drum

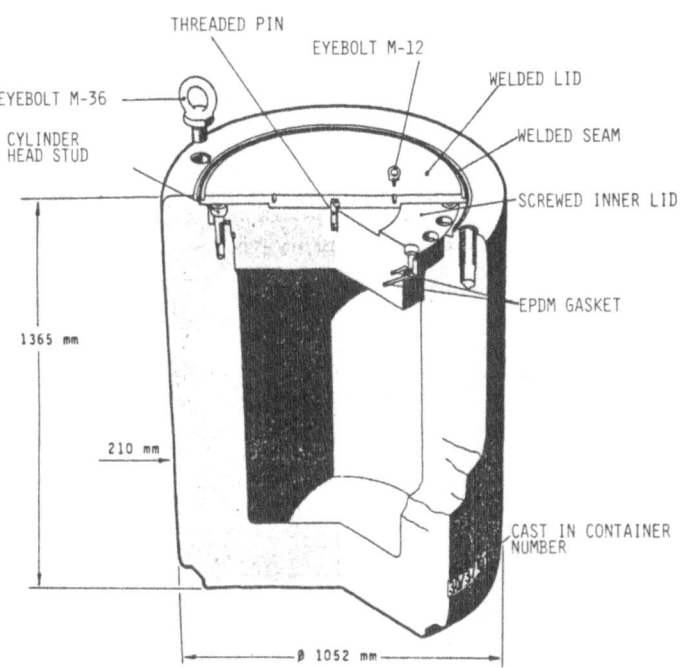

Fig. 2.3 Sphero Cast Iron Container

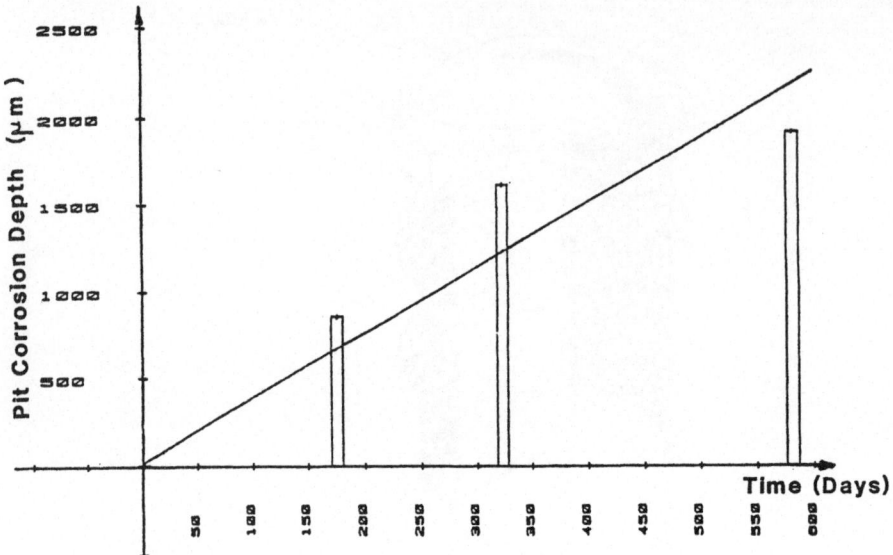

Fig. 2.4   Pit Inner Corrosion Depth of a Sphero Cast Iron Container
(GGG-40-3-Specimen.Waste:Dewatered mixed ion exchange beads
BO_3/Li form.T=55°C)

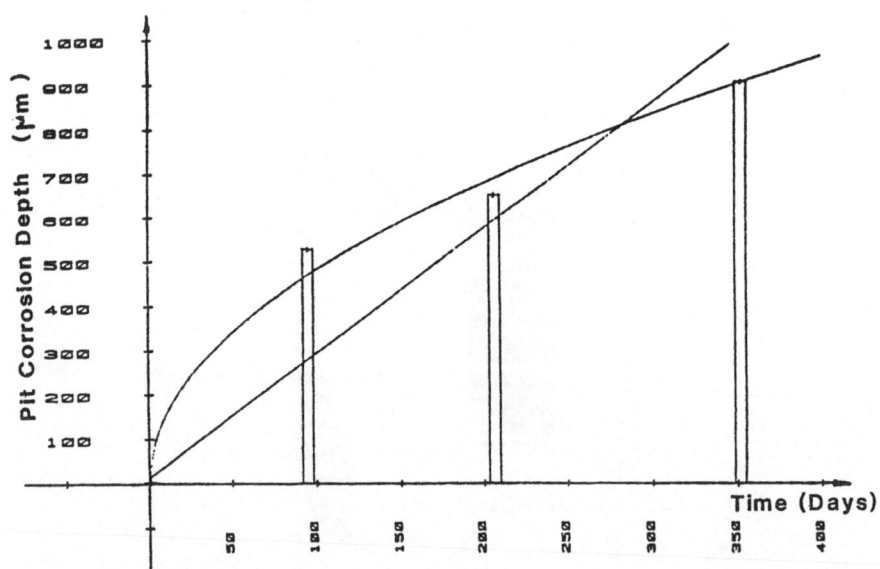

Fig. 2.5   Pit Outer Corrosion Depth of a Sphero Cast Iron Container
(Disposal in a sealed KONRAD-Section.T=42°C;Relative
humidity=85%)

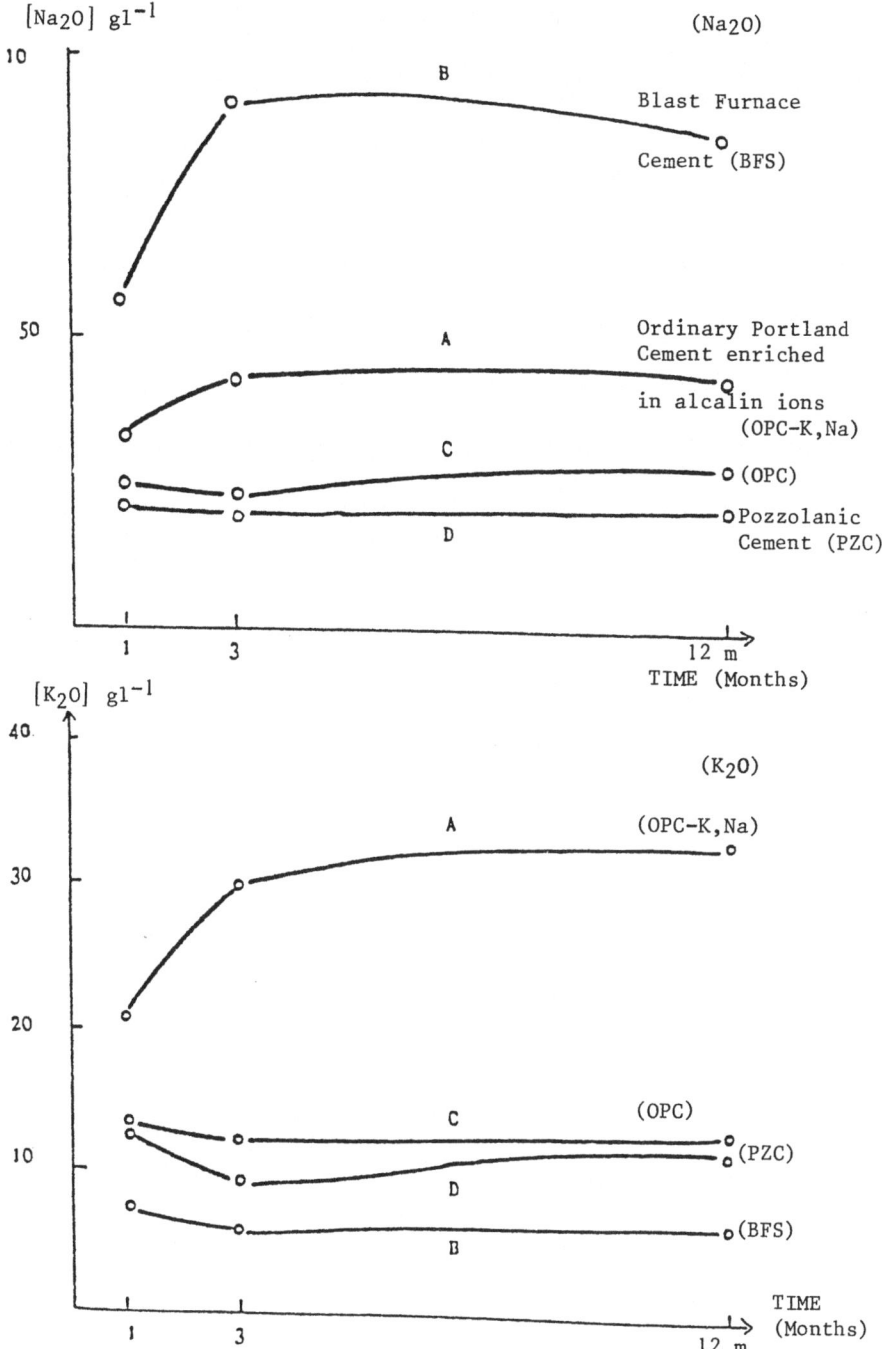

Fig. 2.6   Na20 and K20 Variation in Interstitial
Water for Different Cements

SESSION 3

PRESENT SITUATION AND MAIN RESULTS OF WASTE FORM
CHARACTERIZATION (BEHAVIOUR IN AQUEOUS ENVIRONMENT)
AND DIFFICULTIES ENCOUNTERED WHERE USING CHARAC-
TERIZATION DATA FOR RISK ANALYSIS ASSESSMENT

SYNTHESIS OF SESSION 3

Chairman : Mr J. CLAES - ONDRAF/NIRAS

Subject of discussion : Present situation and main results of waste
form characterization (behaviour in aqueous
environment), and difficulties encountered
where using characterization data for risk
analysis assessment

Objective : Applicability of leaching test data.

Scope : Available leach testing methods and principal parameters
controlling leach test.

SUMMARY OF DISCUSSION

3.1 - Status of waste characterization

Leach resistance is one of the most important properties of a waste
form.

Leach testing can be performed with different objectives in mind :

- Acceptance of packages to the repository.
- Comparison of matrices.
- Prediction of long term behaviour.

The method of testing and reporting of the results depends on the
objective ; choices have to be made on :

- the method of leaching test (static, dynamic),
- the specimen size (laboratory or full scale),
- the waste form type (inactive, active, or simulated waste form...),
- the leachant (demineralized, standard, site specific water...),
- the presentation of the results,

and the parameters relating to the leaching system (time, temperature,
pressure, sample surface/leachate volume), to the leachant
(composition, flow rate, renewal frequency, pH, oxidizing conditions)
and to the waste product (composition, preparation, surface condition
and porosity) have to be defined. In the case of leaching under
disposal conditions, additional factors must be considered :

- hydrology of the disposal site,
- composition of the aqueous medium,
- knowledge of the water flow patterns in the repository,
- evaluation of the waste package quality and the exposed package
surface,
- radionuclide inventory of the package.

After describing the available leach test methods and investigating the effects of the main parameters on radionuclide release, the difficulties encountered in using leach test data to understand the mechanism of radionuclide release, to predict real leach behaviour of packages or to perform risk studies... will be discussed.

It must be noted that the following discussion is limited to homogeneous wastes solidified with the usual matrices (ion exchange resins, evaporator concentrates, in cements, bitumen or polymers). Homogeneous waste is here supposed to comprise also materials where small waste particles are distributed uniformly in a matrix material. On the micro scale, such materials are not homogeneous and this should be kept in mind when evaluating their behaviour (see also section 3.3.3).

3.2 - Applicability of leach test methods

3.2.1 - Laboratory leach test

3.2.1.1 - ANS 16.1 tests / 3.1 /

The test has been developed by the American Society and is designed to measure the leachability of "non selfheating" solidified wastes, that is, waste forms which in their environment will not significantly exceed the temperature range of this test (17,5°C - 27,5°C), i.e. shallow land burial. It is now standardized, and consists of a procedure in which the leachant is changed after designated intervals (2 and 7 hours after the test starts, then every 24 hours for four days ; Beyond this period, intervals between leachant renewals are 14, 28 and 43 days, i.e. a total test duration of 90 days). A short term test lasting for one week can also be used.

The data obtained using this standard are expressed as a material parameter of the leachability of each specie leached.

This parameter is a figure-of-merit called "Leachability Index" (LI). It has a specific meaning for each solid tested. It expresses the leaching data in terms of mass-transport theory, but without implying that the long term leaching mechanisms are known.

This index can thus be used as a basis for comparison of the radionuclide retention capability of various solidification matrix materials. In addition, as this parameter is a function of actual leaching behaviour, it can be applied to the modelling of isotope migration in shallow land burial. On a plant scale, the leachability index can be employed as a means of demonstrating that waste forms meet minimum quality requirements. As a result, the Nuclear Regulatory Commission has issued recommended minimum leachability index specifications in support of 10 CFR 61 waste form stability requirement (LI > 6).

3.2.1.2 - Equilibrium leach test

At Harwell, UKAEA, the equilibrium leach test (ELT), in which solubilities are established, has been selected for laboratory study of the radionuclide release. It approximates to disposal site conditions where groundwater has a very low flow rate : results of this test should represent the maximum levels that nuclides can reach in groundwater under minimal flow conditions.

The equipment used is described in Fig. 3.1. : the individual containers are polythene bottles, 400 ml in volume, fitted with rubber seals and screw on lids.

Into each lid is sealed $\sim$ 1 metre of 1-2 mm bore polythene tube to lead away any gas formed by radiolytic and corrosion reactions. Each filled bottle, with tube, is placed inside a steel outer container. The lids of the outer containers are punctured, but below the aperture a porous bag of Sofnolite (an absorbent for carbon dioxide) is secured, so that any air entering the container has its carbon dioxide removed.

One application of this test is the measurement of equilibrium concentrations of radionuclides in the presence of repository constituents ; one example is given here : it deals with the determination of equilibrium values for uranium, plutonium, neptunium and americium and for this purpose, two forms of sample were tested : waste immobilized in the matrix (cement or polymer) and waste mixed with crushed matrix.

Solubilities of U, Pu, Np and Am in the presence of different anions (SO_4^{2-}, CO_3^{2-}) and also their degrees of sorption on cement have been determined (see table III.1 and III.2). On the other hand, theorical equilibrium concentrations of Cs, Sr, Np, Am and Pu have been calculated and a comparison of the calculated and experimentally determined values is given in table III.3. The agreement between the two sets of data is encouragingly close, the results for each isotope agree to within an order of magnitude. The agreement indicates that :

- the average values obtained with Cs, Sr, Np, Am, Pu can indeed be regarded as equilibrium values,

- sorption is important in determining equilibrium values, and this has kept the solution levels of plutonium, americium and neptunium below their solubility limits.

The measured equilibrium solubilities were compared with the reference drinking water level. The equilibrium value for uranium is below the reference level, while that for neptunium is only a factor 2-3 higher. The values for the long lived isotopes of americium and plutonium are about 2 orders of magnitude higher than the reference drinking water levels (see table III.4).

This equilibrium leach test has the advantage of being precise, reproducible and applicable to the simulation of all types of repository conditions. The test duration is mostly in the range of 90-270 days, but results indicate that equilibrium can be established at shorter times (the equilibrium leach test is also the subject of / 3.2 /).

3.2.1.3 - Wet and dry cycle leach test

Radionuclide release from low level waste forms subjected to wet/dry cycles, simulating burial in the unsaturated zone, is investigated at Brookhaven National Laboratory (BNL). The objectives of this investigation are to provide leach data which will assist the Nuclear Regulatory Commission (NRC) in evaluating the acceptability of LLW Forms for disposal, and to provide realistic source term data for modeling radionuclide transport at LLW disposal sites. (The definition of LLW, for the United States where the "wet and dry cycle test" is developed is given § 4.1.1).

The cyclic wet/dry leach testing of waste forms involved the following steps :

Briefly, a test specimen is placed in a porous medium contained in a column. In order to minimize the sorption of leached radionuclides, an inert material (high density polyethylene (PE) beads) was selected for porous medium. The specimen is surrounded on all sides by a 2-inch-thick layer of PE beads.

Deionized water is used as the leachant, with the total volume of leachant being 10 times the exposed surface area of the specimen during the wet period. The wet period of the total wet/dry cycle is kept constant at one day, while the "dry" period varies ; dry period is defined as the period during which no water is added to the bead column. All experiments are conducted at ambient room temperature.

Investigations have been performed with a simulated waste form consisting of cation exchange resin (IRN-77), loaded with Cs-137 and Sr-85 as tracers, and solidified in Portland type I cement. The test specimens are cylinders of nominal diameter and height of 5 x 10 cm respectively.

Results of cesium and strontium release are presented in Figs 3.2 to 3.4, and show that the processes occuring during the unsaturated period of the wet/dry cycle can have a profound effect on the overall release of radionuclides, especially strontium. Significantly lower releases of Sr-85 were observed compared with Cs-137 from resin/cement waste forms subjected to cyclic wet/dry leach conditions. Cesium release when considered as a function of the actual immersion time (total length of the wet periods of wet/dry cycles) remains relatively constant, irrespective of the varying lengths of the dry periods (Fig. 3.2.b).

The observed cesium release is also comparable to that based on an IAEA-Type leach test, which represents continuously saturated leach conditions (the modified IAEA test corresponds to a dry period of zero).

In contrast, the strontium release is significantly lower and exhibits a systematic decrease with increasing length of dry period (Fig. 3.3.b). This is also reflected in the effective diffusivity values, which vary from $\sim 10^{-9}$ cm^2/s in the one-day dry period experiment to $\sim 10^{-14}$ cm^2/s in the six-day period experiment. Cesium, on the other hand, has an effective diffusivity on the order of $\sim 10^{-8}$ cm^2/s, regardless of the length of the dry period in the cyclic leaching experiments.

The observed differences in their release behaviour can be attributed to different mechanisms by which these radionuclides are mobilized and transported from the bulk matrix into the leachant. In the case of cesium, which is a highly soluble ion, diffusion is the principal release mechanism. Ion-exchange of cesium from the resin and dissolution of cesium compounds are too rapid to be rate-controlling ; strontium, on the other hand, is relatively insoluble and may even become part of the cement gel matrix during curing of the waste form. Consequently, release of strontium involves concentration levels (common ion effect) in the pore water and leaching of the matrix and subsequent diffusion of the leached strontium into the leachant. Because of additional rate-controlling, dissolution steps, the observed release rate of strontium is much lower than that of cesium.

Decreasing amounts of strontium release with increasing length of "dry" periods may be due to progressive tortuosity and decrease in pore volume as the cement continues to hydrate, even during the "dry" period. (Decreasing amounts of Sr-release with increase length of dry period may also associated with the redistribution of strontium between the resin and the cement matrix during the curing process).

In order to bound the uncertainties in radionuclide release from a given waste form, it is also important to determine the two following types of release limits :

(a) The solubility limiting release under static or very low flow rate conditions.

(b) The leach rate-limiting release under high flow conditions.

The resultant upper and lower bounds will provide estimates of the release under a range of flow conditions.

The normalized cumulative fractional releases of Cs-137 and Sr-85 (and leach rates) obtained with this two types of test are represented in Fig. 3.5 (see also / 3.3 /).

The behaviour of other radionuclides is currently being examined.

3.2.1.4 - IAEA test performed at RISØ

To obtain a situation relating to a one dimensional diffusion model and to minimize the effects of sample preparation, an IAEA modified test is developed at RISØ to study the leach behaviour of cement waste form with this following particularity : specimens (cement containing sodium sulfate and doped with Cs-134 and Sr-85) are prepared in polyethylene bottles, which are also used for the leaching test. When the cement has hardened, the bottle wall is squeezed onto the specimen circumference in order to obtain a one dimensional geometry with no leaching from gaps between sample and container. In this way, the leached thickness is a direct measure of the leach rate. With this very simple set-up, allowing to work with a constant cross section, series of tests under a variety of conditions and with various parameters have been carried out.

In a series, Sr-85 and Cs-134 releases from sodium sulfate solidified in Italian Pozzolan cement (IPOZZ) were studied for two types of water, with and without access of CO_2 from the atmosphere and for 2 rates of sampling. One example of a series is given schematically in Fig. 3.6. Results are presented as plotted versus the square root of the time, as shown in Fig. 3.7. A significant reduction of strontium release, due to the absorption of atmospheric CO_2 is indicated in Fig. 3.7.b.

3.2.2 - Full scale leaching test

3.2.2.1 - Full scale leach testing at CEA-SACLAY centre

Full scale leach tests are conducted in the CEA-SACLAY centre with real packages ; the tested matrices are based on cement, bitumen, polymers and mixtures of these ones ; package volumes are in the range 10 - 900 liters.

The objectives of these experiments are : package characterization, quality control and quality assurance, R and D studies (scale effect - see session 5 § 1.1.).

The activity release is monitored by measuring Pu-238, Am-241, Cs-137, Sr-90 and Co-60. in addition, matrix constituents like Na, K, Ca, Cl are measured too. Subsequent to the tests, some of which extended over a period of four years, an attempt was made to establish the activity balance as accurate as possible :

Activity balance

The released activity is not only present in the leachate (A_1). It is also present in any sludges or detritus produced (A_s), on the filters (A_f) and in the pipework of the loop (A_p).

The radiochemical balance must be :

$$Ao - A'_o = A_1 + A_s + A_f + A_p ,$$

where Ao is the waste form activity before leaching and A_o' after leaching.

In the case of PWR borate/cement waste form (table III.5) the majority of the Cs-137 is dissolved in the leachate, while the bulk of the Sr-90 has plated out in the loop.

In the case of the alpha emetters, the bulk has also plated out and there is general contamination in the loop, the sludges and filter as can be seen, for instance, for Pu-239 release from bitumen/Na NO_3 waste form (table III.5). Only a small fraction of released alpha activity is contained in the leachate.

Complementary examination

To complement these full scale leach experiments, examinations are carried out after leaching. In particular, continuous gamma spectrography (GeLi detector-collimator of 20 and 2 mm) is performed on core samples taken from the blocks to reveal the internal distribution profiles of Cs-137, Co-60 (see Fig. 3.8). These gamma scans provide an estimate of the tested waste form homogeneity ; the quantities of released radionuclides can be calculated and in this way, compared with the cumulative activity releases obtained by leachate analysis. A factor of 3 is observed between these 2 values (for more details, see Ref. / 3.4 /).

3.2.2.2 - Full scale experiment at Brookhaven National Laboratory

Full scale leaching experiments are also performed at BNL / 3.5, 3.6 / to study the release of radionuclides from actual commercial reactor wastes solidified in masonry cement, portland type III cement and vinyl ester-styrene (Dow polymer) (table III.6). The tests are conducted under standard leach conditions (ANS 16.1 standard) ; Data are plotted versus cumulative leaching time for a total of 316 days (see Fig. 3.9 to 3.12).

It is observed that Co-60 releases are always lower than those of Cs-134 or Cs-137. In the case of cement matrices, the extremely low leachability of Co-60 relative to Cs-137 is attributed to its chemical interactions within the matrix.

For BWR waste forms, the Co-60 cumulative fraction release is approximately 3 times greater for the Dow than for the equivalent cement waste form (Fig. 3.12). On the other hand, the cesium releases from BWR waste are approximately one order of magnitude greater for cement than for Dow (Fig. 3.10).

Shallow land burial in situ lysimeter tests are also being performed at Savannah River Laboratory and Battelle Pacific Northwest Laboratory in conjonction with BNL to compare releases in these cnditions and releases under standard leach tests conditions / 3.7 / ; In the in situ tests, the isotope migration is monitored, and it appears that Co-60 is migrating rapidly through the soil ; (studies going on in CEA on the formation of complexes indicate that cobalt has a tendency to form complexes in soils, which are liable to replace the calcium/soil complexes). Fig. 3.13 is a schematic of the in-situ lysimeter.

3.3 - Discussion

Leach tests are being performed to obtain data for package characterization, for risk analysis assessment and to predict the long term behaviour of disposed waste forms. Some questions arise :

- Waste type
The leach tests which are described in this session concern liquid and/or particulate wastes uniformly dispersed in a matrix -the so-called homogeneous waste forms. What leach tests can be applied to immobilised heterogeneous wastes, where the activity may still be locked up in waste or simply as surface contamination ?

- Representativity of sample
Do tracered simulates and real wastes have the same leaching behaviour ?

- Specimen size
Can data obtained with laboratory scale samples be applied to full scale packages.

- Ageing effects on waste form leaching
What are they, and how can they be simulated without changing the release mechanisms ?

- System parameters
What are the effects of temperature, pressure, the frequency of leachant renewal, the nature of the leachant... on the leaching behaviour ? Is it possible to extend the data obtained from laboratory tests to real burial conditions ?

- Test duration
Are short term tests adequate, especially for long term prediction ?

- Leachate analysis and the mode of reporting

3.3.1 - Heterogenous wastes

Heterogeneous solid wastes cannot be neglected because they may contain a considerable fraction of long-lived radionuclides, especially in the case of reprocessing wastes. As an example, the heterogeneous solid LLW from reprocessing would correspond to about 10 % of the alpha-activity for a land disposal scenario of LLW and MLW as evaluated for the Belgian nuclear programme. Therefore, the knowledge of the behaviour of the packages containing these wastes is important. Because of the macro-scale heterogeneity, small size experiments are unlikely to be practicable and even for large scale experiments, the representativity is of major concern.

3.3.2 - Representativity of sample

Comparative leaching tests over 28 days and 3 months have been performed at BNL with samples containing real wastes and tracered samples. The comparison of Cs-137 release behaviour indicates a lower release from real reactor waste than from the simulated one, resulting in 3 or 4 fold reduction in its effective diffusivity value. The leach behaviour of both waste forms exhibits considerable similarity, hence it appears reasonable to assume that the Cs-137 leach rate derived from testing of simulated waste composites can be employed to evaluate and predict the release behaviour from reactor wastes, with the proviso that the specimens are prepared and leached under identical conditions / 3.8 /.

3.3.3 - Sample size

Two aspects have to be examined : Can data obtained from laboratory samples be extended to real drums and are these results significant ? The first aspect is discussed in the fifth session ; for the second aspect, a French approach is as follows : the leach rate, expressed in term of equivalent leached mass, is compared to the boundary mass "H" for which the waste form is considered homogeneous (relative to the activity) ; H is defined as following : a waste form will be considered homogeneous at H-Mass level, if the standard deviation $\frac{\sigma AH}{AH}$ of the mean activity AH for a sample of H-Mass is smaller thant 50 %.

The comparisons between the equivalent leached mass calculated for each leaching sequence and H, for small samples and full scale IER/polymer specimens indicate that for small samples, the equivalent leached mass is principally below the homogeneity boundary ; thereby leach results are not significant (Fig. 3.14).

On the other hand, full scale data are above the homogeneity boundary and have to be taken into account (Fig. 3.15).

3.3.4 - <u>Ageing effects on waste form leaching</u>

The first question in : how does ageing affect the leaching characteristics ?

This point is being studied at Harwell. Ageing phenomena are partially induced by radiation, consequently the impact of irradiation effects have to be investigated. For this purpose, different waste forms are used :

- Zeolite in cement : Inactive and active specimens (the latter are doped with Cs-137 and Sr-85 to obtain an activity level of 15 to 20 mCi/l of waste form) were stored for two years, some of them in an irradiation field (approximately 1 MRad/h) prior to leaching. It was observed that the pozzolanic reaction is speeded up in an irradiation field. However, the leach rate for Cs and Sr were marginally reduced.

- Magnox hull debris in cement : cured for 90 days before leach testing, the magnox and any associated fuel tend to corrode even in the high pH that exists in cement. This might be expected to lead to finite leach rates for U and Pu. However, neither was detected after 28 days of leaching. Clearly a much longer ageing period must be allowed before any information relevant to long term leach rates is meaningful.

Since MLW in the UK is likely to be stored for several tens of years prior to disposal, drying of the waste form also has to be anticipated the assessment of ageing effects.

- Magnox fuel pond sludges in bitumen : inactive and active samples (doped with Cs-137 and Sr-85 to obtain an activity level of 10 mCi/l were stored for 2 years, some of them in a radiation field, prior to leaching. The irradiated samples showed distinct changes in microstructure and a tendency to swell due to gas bubble formation depending on the ratio of specimen size to irradiation dose.

- PWR evaporator concentrates in vinyl polyester : in the case of externally irradiated small samples, sodium nitrate was squeezed out of the matrix ; however, this phenomenon was neither observed with larger specimens nor with epoxy resins. In general, leach rates increased in a radiation field.

These four examples demonstrate that irradiation effects cannot be generalized, each waste form showing a different behaviour.

Ageing tests on bituminized EUROCHEMIC sludges (Eurobitumen) are also underway in Belgium. To date, after one year storage of the samples, no change in physical properties have been observed.

A second point relative to the ageing effect concerns the possibility of conducting accelerated leaching tests :

– Generally, during ageing, the microstructure of the material changes ; but these changes might be different if the process is accelerated : for instance, using high dose rates to accelerate the effect of ageing can lead to coke formation in bitumen ; accelerating ageing by crushing waste form also leads to modification of the microstructure : with such a test, (comparison of Cs release from monolithic and crushed waste form), it was observed that Cs release decreases with increasing the specific surface area ; this phenomenon is attributed to a change of the surface state (surface oxidation). Comparative tests, carried out at KFK on accelerated corroded and uncorroded samples of cement/NaNO$_3$ (10 %) lead to leach rates a factor 10 higher for the corroded samples. Another way, in which the acceleration of ageing also modifies a system is the large difference observed between bitumen samples irradiated at low and at high dose rates. It seems that only low dose rate irradiation leads to leach rates similar to those in unirradiated samples.

Another cautionary example of accelerated testing is illustrated by diffusion experiments conducted at Riso. Test carried out in a diffusion cell (Fig. 3.16) showed that the Cl-36 diffusion through a concrete membrane is time dependent, which, is not taken into account in an accelerated test (see § 5.2.2.1).

It is suggested that non destructive monitoring of the elastic modulus of a waste form may be a way of assessing changes due to natural or artificial ageing.

3.3.5 – Influence of system parameters on the leaching behaviour

Pressure

Pressurization effects (if any ?) are dependent on the sample porosity ; in the case of samples from the bituminization process of ILLW at EUROCHEMIC, no significant pressure effects were observed up to 230 bar.

Temperature

Effects of temperature should not be generalized : for instance leaching tests performed at RISØ (similar to the tests described in 3.2.1.4), on sodium sulphate/Portland cement waste forms, (previously stored for 216 days in water at 20 or 40°C) with the two same leaching temperatures (20 and 40°C) indicate higher values for the static test at 40°C, (factor of two between the diffusivities obtained at this two temperatures), as would be expected for leaching controlled by diffusion water in the pore system in concrete (see Fig. 3.17). The difference is, for modelling purposes, of minor significance. For bitumen, temperature represents a very important parameter, as can be seen in table III.7, for Cs-leachng in different media (demineralized water-DM and clay-water-mixture-CWM) for two temperature : 23 and 40°C. A complicating phenomen is that, i.e, in the case of demineralized water, the swelling increases from 99 % ($\frac{\Delta V}{V}$) for 23°C to 890 % for 40°C. Changes in temperature can inhance phenomena (like swelling) rendering even more difficult the interpretation of the results.

Frequency of leachant renewal

It has been pointed out that changes of leachant, as indicated by the results from various laboratories, interrupt the build up of the specific environmental chemistry.

Nature of leachant

The effect of leachant composition cannot be generalized, it depends on the type of waste form.

For instance, small samples from the bituminization process of ILLW at EUROCHEMIC containing 25 % sodium nitrate, releases all the nitrate after two years of leaching in a clay water mixture at 23°C (fig. 3.18) ; (the corresponding volume change amounts to 100 % ; see Fig. 3.19) ; this release is more prononced in clay water mixture than in demineralized water.

The following example shows also, the influence of leachant composition : a leach test series was conducted with sodium sulphate/cement specimens. Some were stored statically in quinary-brine before leaching in NaCl solution, which was replaced and sampled with a pre-determined frequency. Some were stored statically in NaCl before leaching in NaCl. In the case of storage in Q-brine, the amount of Cs leached during static storage was considerably smaller than in the case of storage in NaCl. The probable reason is pore blocking in the sample surface by precipitation of Mg-containing minerals, which renders it less permeable to dissolved species. However, the effects seem to be only temporary since on changing to leaching in 5M NaCl, there is a considerable increase in leach rate on the sample initially stored in Q-brine (see Fig. 3.20).

3.3.6 - Test duration

Fig. 3.21 and 3.22 show the cumulative fraction of Cs and Co released as a function of time, for BWR evaporator concentrate plus ion exchange resin waste immobilised in cement. After 120 days the Cs and Co leach rates suddenly increased, suggesting failure of the sample. Due to the time scale effect, these unpredictable increases would not have been observed in a short term test. This example shows that a short term test would have been misleading. But supposing a sample does not fail until after 350 days under test, i.e. how short is short ?

3.3.7 - Leachate analysis and mode of reporting

As has been pointed out earlier, the cumulative fraction released or the equivalent depletion depth is an extremely useful way of presenting leach data.

It has become standard practise to plot the cumulative fraction leached as a function of square root of time, since approximation to a straight liner suggests a diffusion mechanism. From this data, an effective diffusivity can be calculated. However, there is very frequently an initial non linear portion of the relationships, which varies in duration in an unpredictable way. So the questions remain : what is the first phase duration ? and has it to be taken into account for risk analysis assessment ?

The release mechanisms are also very sensitive to leachant composition. Currently there is little evidence with MLW forms for colloid formation, but a new and more sensitive speciation method, called laser spectrometry, is being developed at Ispra and at the technische universität- Münich.

3.4 - Session conclusion

It appears necessary for the future to investigate more in detail heterogeneous wastes, to carry out full scale leaching tests and to apply leaching conditions under land burial conditions or in situ experiments. For long term predictions, the basic leaching mechanisms have to be known and the chemical form of leached radionuclide have to be analysed to get informations on mobile/immobilized, solved/precipitated species ; the relevant models are then to be confirmed by long term leach tests. It also appears that prior to embark on accelerated tests, the microstructural changes due to the ageing must be better understood.

REFERENCES

[3.1]   "Measurement of the leachability of solidified low-level radioactive wastes".
American Nuclear Society.
Standards committee.
Working group ANS-16-1 (June 1984).

[3.2]   GREENFIELD B.F et Al.
"Equilibrium leach testing of magnox fuel cladding residues".
AERE R 11390 (1984).

[3.3]   DAYAL R. et Al.
"Wet and dry cycle leaching : aspect of releases in the unsatured zone".
BNL NUREG 3358 (1984).

[3.4]   NOMINE J.C. ; VEJMELKA P.
"Experience with full scale leaching of low and medium level waste".
EUR 8979 (1984).

[3.5]   NEILSON, R.M. Jr ; KALB P.D. and COLOMBO P.
"Lysimeter study of commercial reactor waste forms".
BNL 35 1613, Brookhaven National Laboratory, UPTON, N.Y. 11973 - Septembre 1982.

[3.6]   KALP P.D. ; COLOMBO P.
           "Full scale leaching of commercial reactor waste forms".
           BNL 35561 (septembre 1984).

[3.7]   COLOMBO P.
           "Special waste form lysimeter - Arid site".
           BNL 36547.

[3.8]   ARORA H. - DAYAL R.
           "Solidification and leaching of boric acid and resin LWR
           wastes".
           NUREG CR 3909 (june 1984).

TABLE III.1

THE SOLUBILITY LIMITS FOR U, Pu, Am AND Np IN THE PRESENCE
OF SULPHATE AND CARBONATE IONS

Solution	Concentration (mol dm^{-3})			
	U	Pu	Am	Np
Concrete leachate	1.5×10^{-7}	2×10^{-8}	4×10^{-10}	2×10^{-6}
Saturated Ca(OH)$_2$/ CaCO$_3$	1×10^{-4}	3×10^{-9}	4×10^{-11}	5×10^{-8}
Concrete leachate/ saturated CaCO$_3$	–	–	–	10^{-7}
Saturated Ca(OH)$_2$/ CaSO$_4$	3×10^{-1}	8×10^{-9}	2×10^{-10}	8×10^{-8}
Concrete leachate + saturated CaSO$_4$	4×10^{-7}	1×10^{-8}	3×10^{-10}	1×10^{-7}
Saturated CaCO$_3$	–	–	–	2×10^{-8}
Saturated CaSO$_4$	5×10^{-4}	2×10^{-8}	2×10^{-9}	2×10^{-7}

Nota :
Samples were centrifuged and filtered before analysis.
- Indicates not determined.

TABLE III.2
THE SORPTION OF ACTINIDES ON CRUSHED CEMENT AND POLYMER

Actinide	Concrete : sorption coefficient Kd at aqueous : solid (w/w)			Polymers : sorption coefficient Kd at aqueous : solid (w/w)		
ml leachant	500	100	10	500	100	10
U	–	28	230	–	0	0
Pu	–	2000	8000	–	4	500
Am	–	530	5000	–	130	300
Np	⟶ 1000 ⟵			300	400	500

TABLE III.3
COMPARISON OF CALCULATED AND EXPERIMENTALLY DETERMINED
EQUILIBRIUM CONCENTRATIONS

Nuclide	Cs-135	Sr-90	Np-237	Am-241	Pu-239/40
Concentration	n mol 1^{-1}		p mol 1^{-1}		
Experimental values	–	15	210	5	130
Calculated values*	37	31	40	8	960
Ratio $\dfrac{calc.}{obs.}$	–	2.1	0.2	1.6	7.4

* Based on an average level of Cs-137 = 6×10^{-8} M in the leachates and the following values for the distributions (Kd) of the nuclides between water and cement :

Nuclide	Cs	Sr	Am	Np	Pu
Kd	5	10	5000	1000	8000

TABLE III.4
COMPARISON OF REFERENCE DRINKING WATER LEVELS WITH THE
MEASURED EQUILIBRIUM SOLUBILITIES

Nuclide	U Nat	Np-237	Am-241	Pu239/40
Reference drinking water level* (M)	3×10^{-6}	8×10^{-11}	3×10^{-14}	2×10^{-12} (239) 4×10^{-13} (240)
Measured equilibrium solubility level (M)	0.15×10^{-6}	21×10^{-11}	500×10^{-14}	130×10^{-12}

* To give an adult an annual whole body dose of 0.5 mSv.

TABLE III.5
RELEASED ACTIVITY FOR TWO WASTE FORMS

	Volume (liter)	d (days)	Isotope	A_o Ci	A_1 Ci	A_s Ci	A_p Ci	A_f Ci
Cement/ borate (RWF 2)	110	1.200	^{137}Cs	$6,3.10^{-3}$	$3,3.10^{-4}$	$7,7.10^{-7}$	$1,5.10^{-7}$	7.10^{-6}
			^{90}Sr	$6,3.10^{-3}$	4.10^{-8}	$4,6.10^{-8}$	$4,6.10^{-7}$	
			^{60}Co	?		$(1,9.10^{-8}$	$(1,5.10^{-8}$	1.10^{-7}
Bitumen/ sodium nitrate (RWF 7)	170	?	^{239}Pu		5.10^{-7}	$1,2.10^{-6}$	$1,2.10^{-6}$	
			^{241}Am		$6,9.10^{-8}$	$1,7.10^{-6}$		

TABLE III.6
WASTE TYPE-SOLIDIFICATION AGENT COMBINATIONS EMPLOYED IN
FULL-SCALE LEACHING EXPERIMENT $\lfloor$ 3.5 $\rfloor$

	Solidification agent		
Waste type	Portland Type III Cement	Dow Polymer	Masonry Cement
Sodium sulfate concentrate	BWR[A]	BWR	--
Sodium sulfate concentrate + ion exchange resins	BWR	BWR	--
Boric acid concentrate	--	--	PWR[B]

(A) 800 MWe boiling water reactor.
(B) 830 MWe pressurized water reactor.

TABLE III.7
SAMPLES OF EUROBITUMEN : EFFECTS OF LEACHANT AND TEMPERATURE
(90 days of leaching)

T(°C)	Medium	$\frac{\Delta V}{Vo}$ (%)	NO_3 (%)	U %	$\Sigma\alpha$ %	$\Sigma\beta$ %	Sr %	(28d) Cs %	Co %
23	DW	99	Leachate 35		0.001	2.6	2.5	0.01	1.1
			Total 44	<0.01	0.004	3.5	3.5	0.02	1.3
	CWM	76	Leachate 32	<0.01	n.m.	0.7	0.7	0.001	0.1
			Total 53	0.023	0.04	6.7	2.2	0.01	2.1
40	DW	890	Leachate 56	0.3	0.002	3.8	2.0	0.3	1.1
	CWM	540	Leachate 51	<0.25	0.002	1.8	3.4	0.01	0.6
			Total 70	0.6	0.47	4.9	5.9	0.12	1.9

Scale   1/1

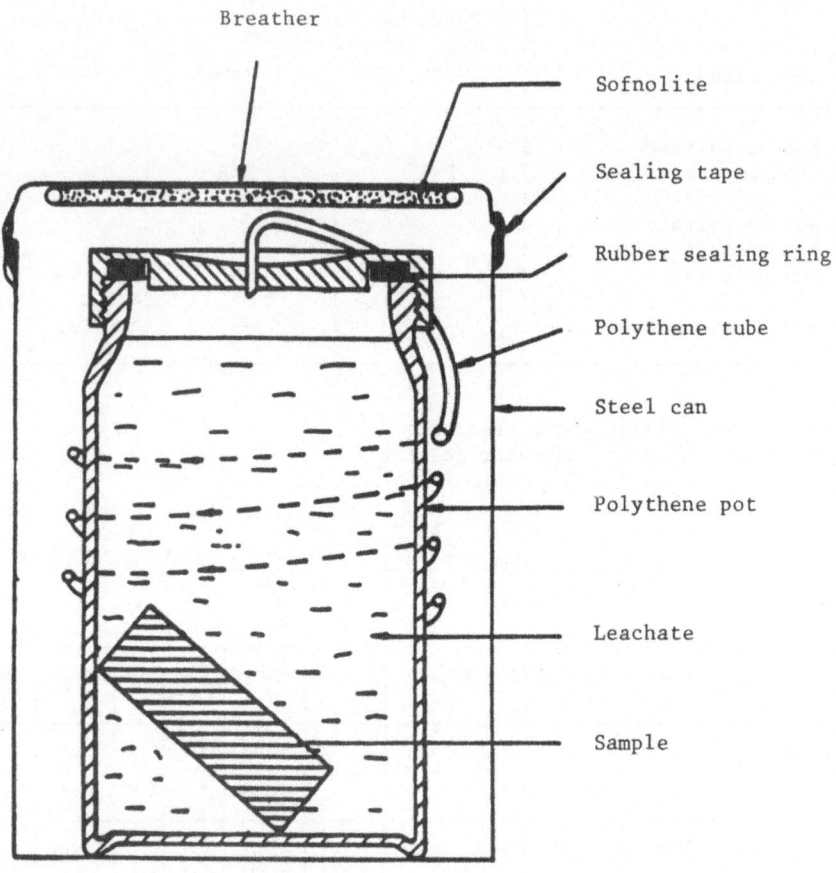

Breather

Sofnolite

Sealing tape

Rubber sealing ring

Polythene tube

Steel can

Polythene pot

Leachate

Sample

Fig. 3.1 Section of a Leaching Experiment

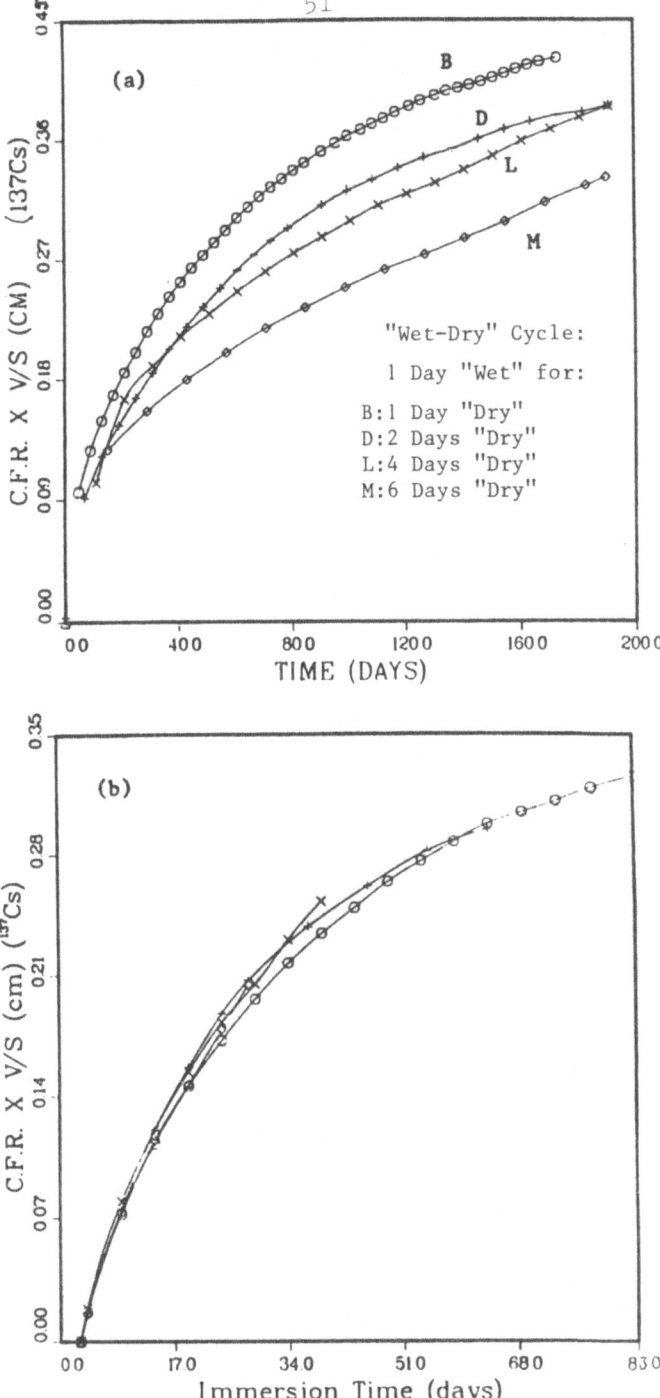

Fig. 3.2 Normalized Cumulative Fractional Releases of
Cs-137 in Experiments B, D, L, and M as
a Function of : (a) Total Elapsed Leach Time
(b) Total Immersion Time

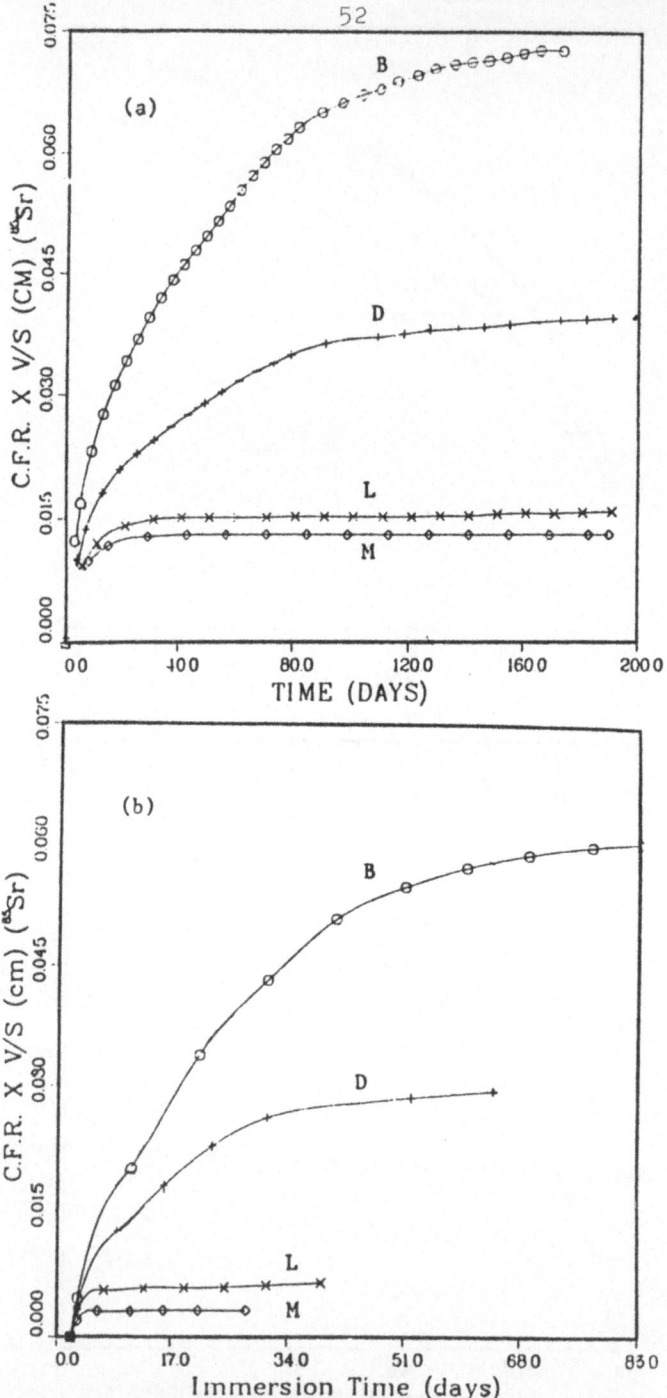

Fig. 3.3 Normalized Cumulative Fractional Releases of
Sr-85 in Experiments B, D, L, and M as
a Function of : (a) Total Elapsed Leach Time
(b) Total Immersion Time

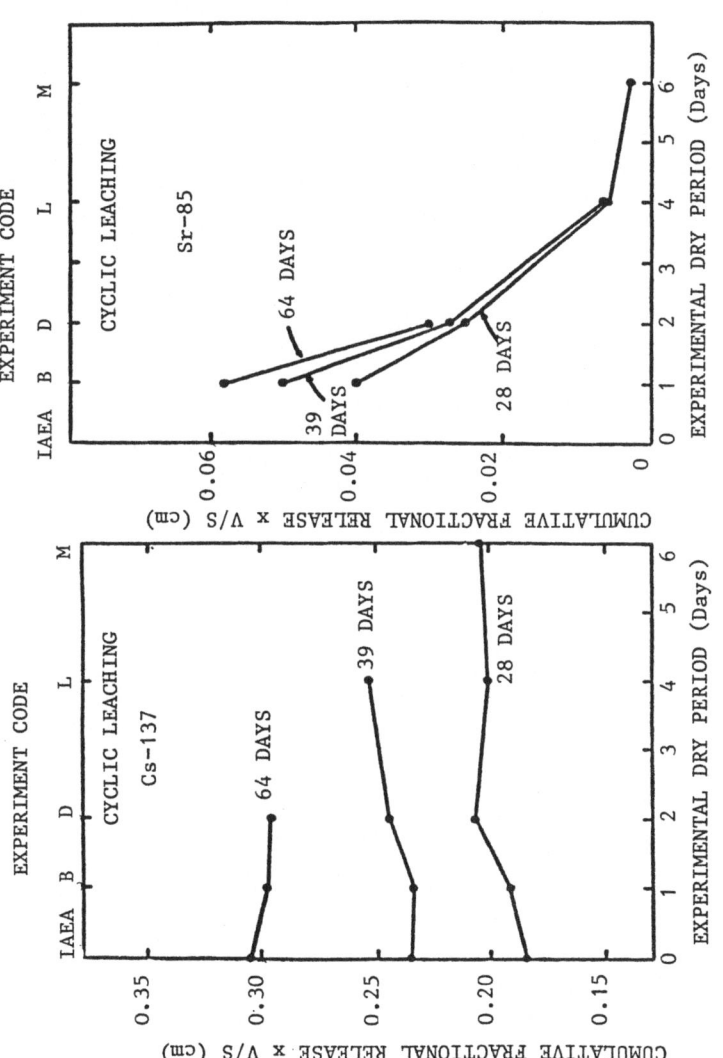

Fig. 3.4 Normalized Cumulative Fractional Releases of Cs-137
and Sr-85 as a Function of Increasing Experimental
Dry Periods in Experiments B, D, L, and M

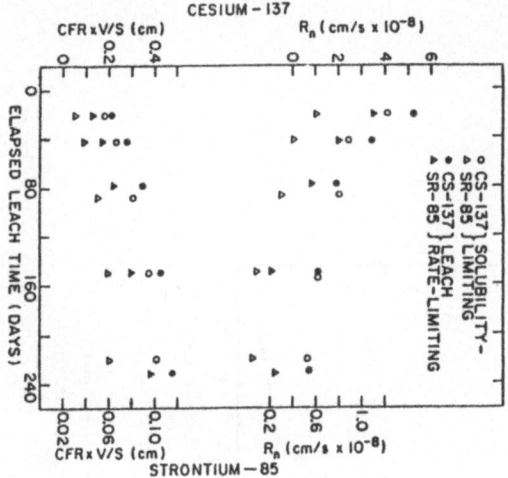

Fig. 3.5

Normalized Cumulative Fractional
Releases for the Solubility &
Leach Rate Limiting Experiments

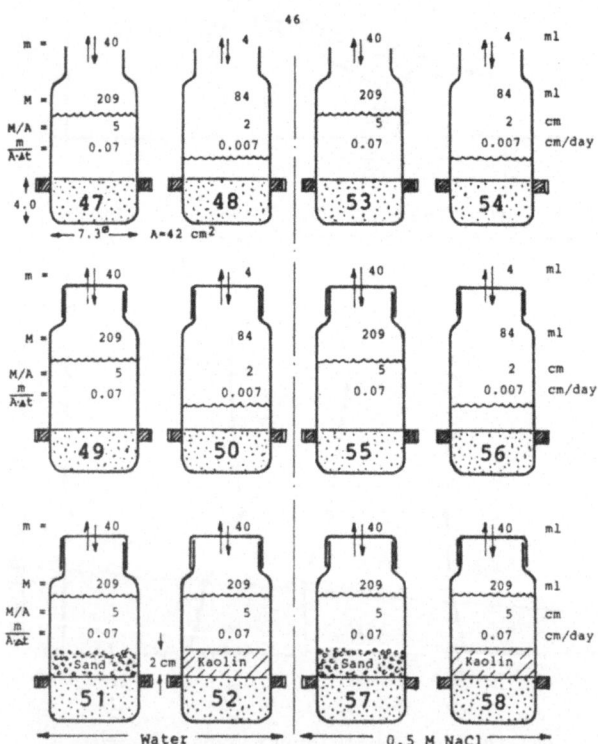

Fig. 3.6 ^{134}CS and ^{85}Sr Leaching from Sodium Sulphate Solidified in
ITALIAN POZZOLAN CEMENT
12 different systems, 2 types of water, with and without
access of CO_2 from the atmoshpere, 2 rates of sampling, etc.
Sampling frequency: Δ t = 14 days

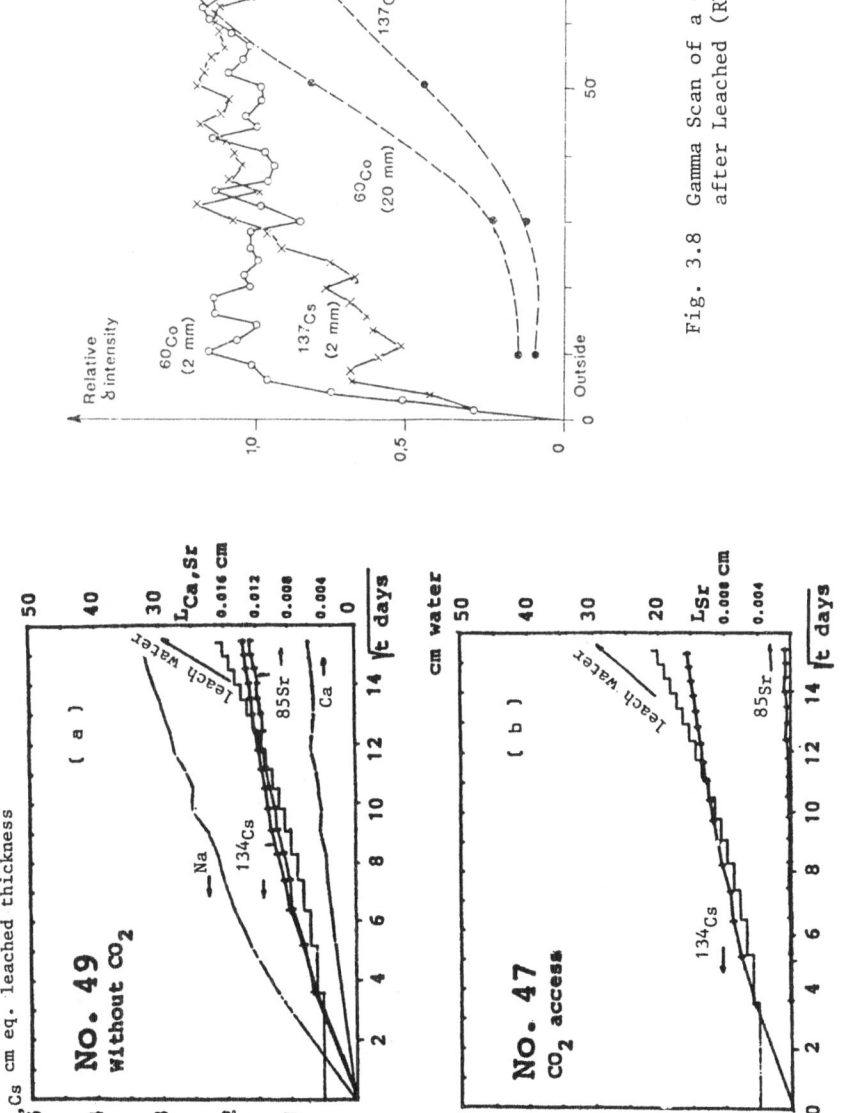

Fig. 3.8   Gamma Scan of a Cement Borate Core
after Leached (RWF 2)

Fig. 3.7   Leaching in Water at 20°C of Sodium Sulphate
Solidified in IPOZZ

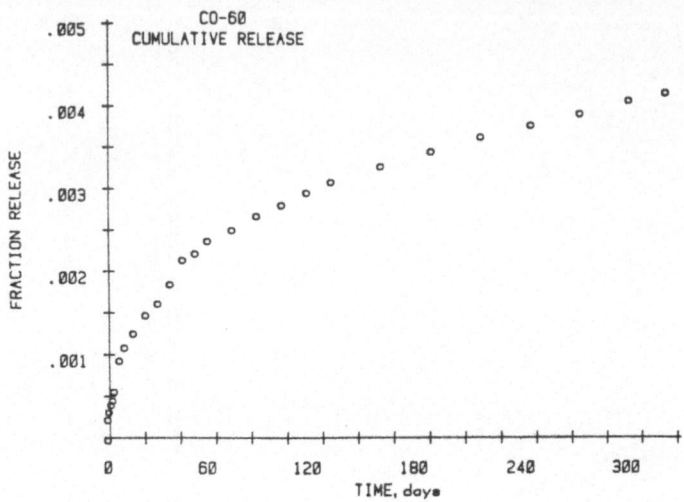

Fig. 3.9   Cumulative Fraction Release of Co-60 as a Function of Time for PWR
Evaporator Concentrate Waste Solidified in Masonry Cement

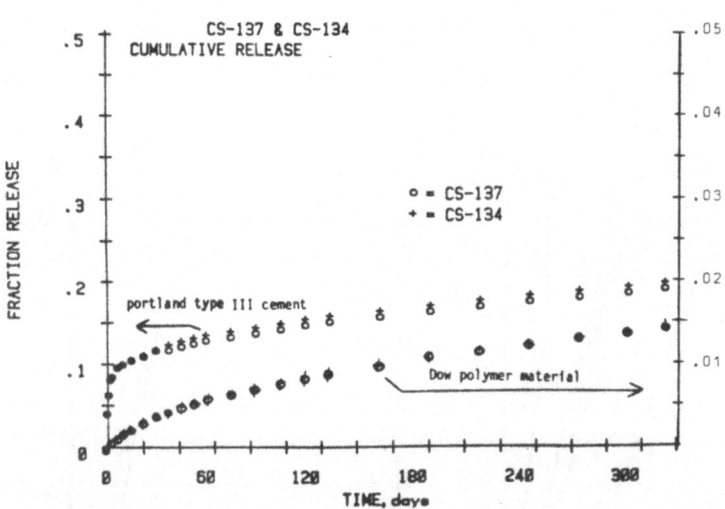

Fig. 3.10   Cumulative Fraction Release of Cs-137 and Cs-134 as a Function
of Time for BWR Evaporator Concentrate Waste Solidified in
Portland Type III Cement and in Dow Polymer Material

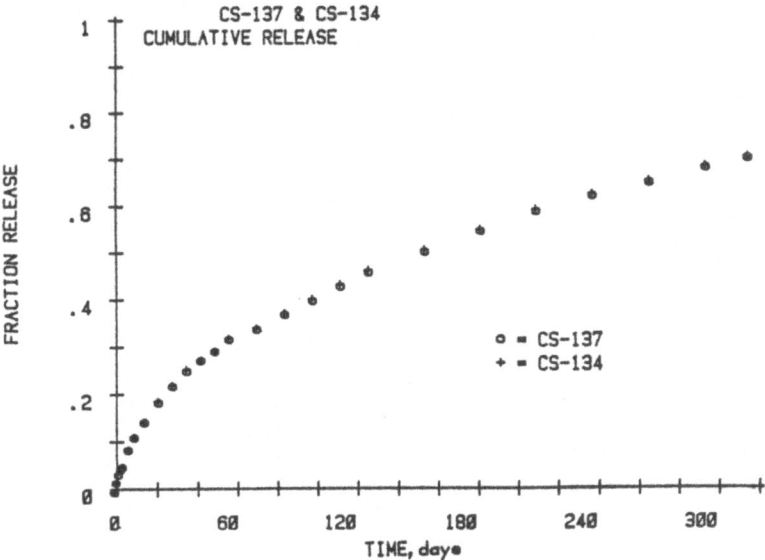

Fig. 3.11   Cumulative Fraction Release of Cs-137 and Cs-134 as a Function
of Time for PWR Evaporator Concentrate Waste Solidified in
Masonry Cement

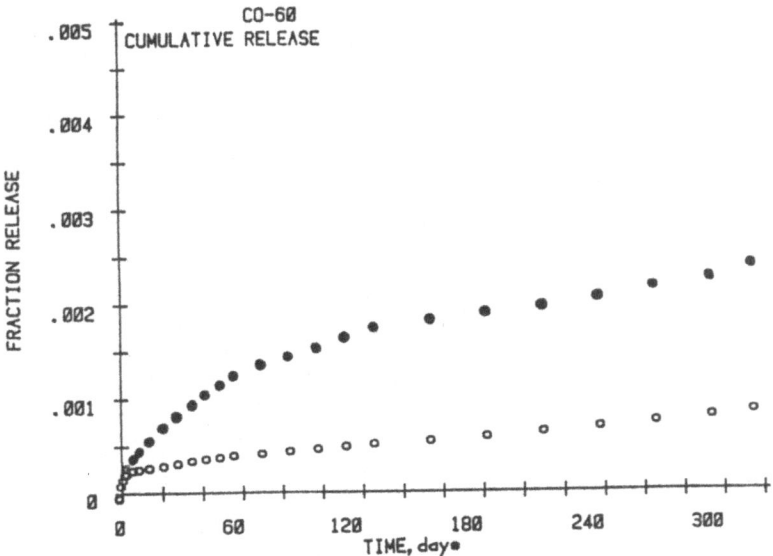

Fig. 3.12   Cumulative Fraction Release of Cs-137 and Cs-134 as a Function
of Time for BWR Evaporator Concentrate Waste Solidified in
Portland Type III Cement and in Dow Polymer Material

58

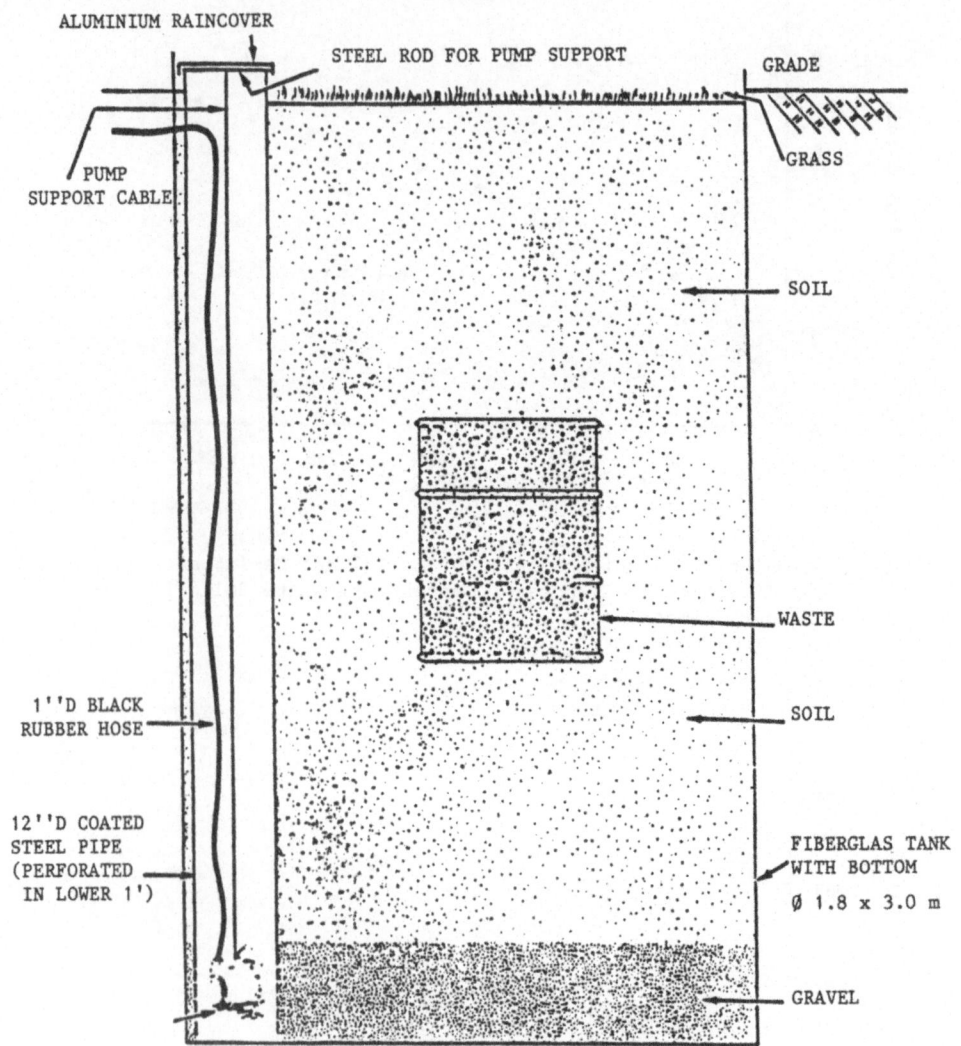

Fig. 3.13 Cut - Away View of a Special Waste Form Lysimeter

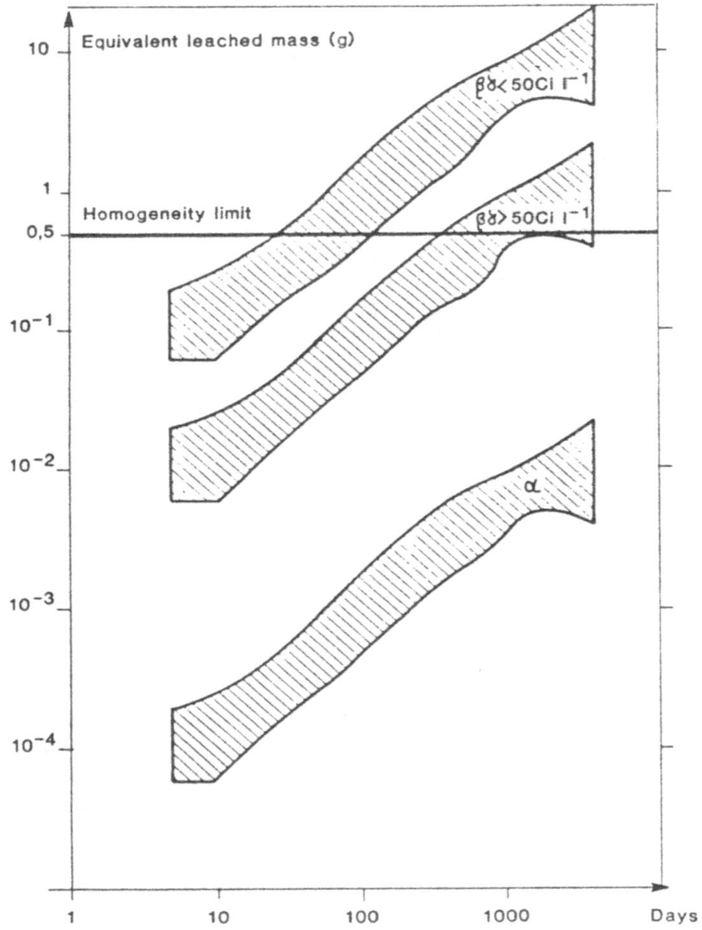

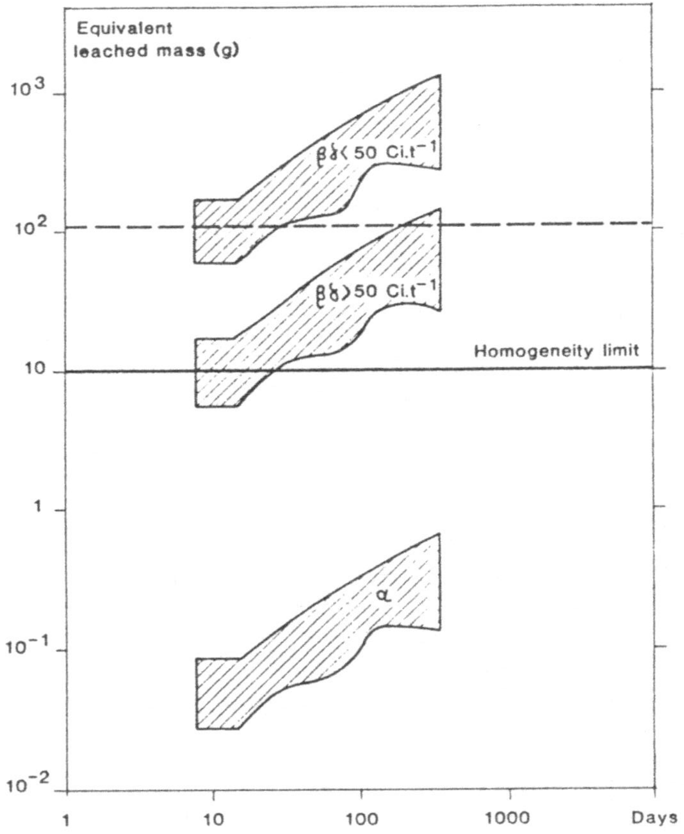

Fig. 3.15 Equivalent Leached Mass
Full Scale Specimen

Fig. 3.14 Equivalent Leached Mass
Small Scale Sample

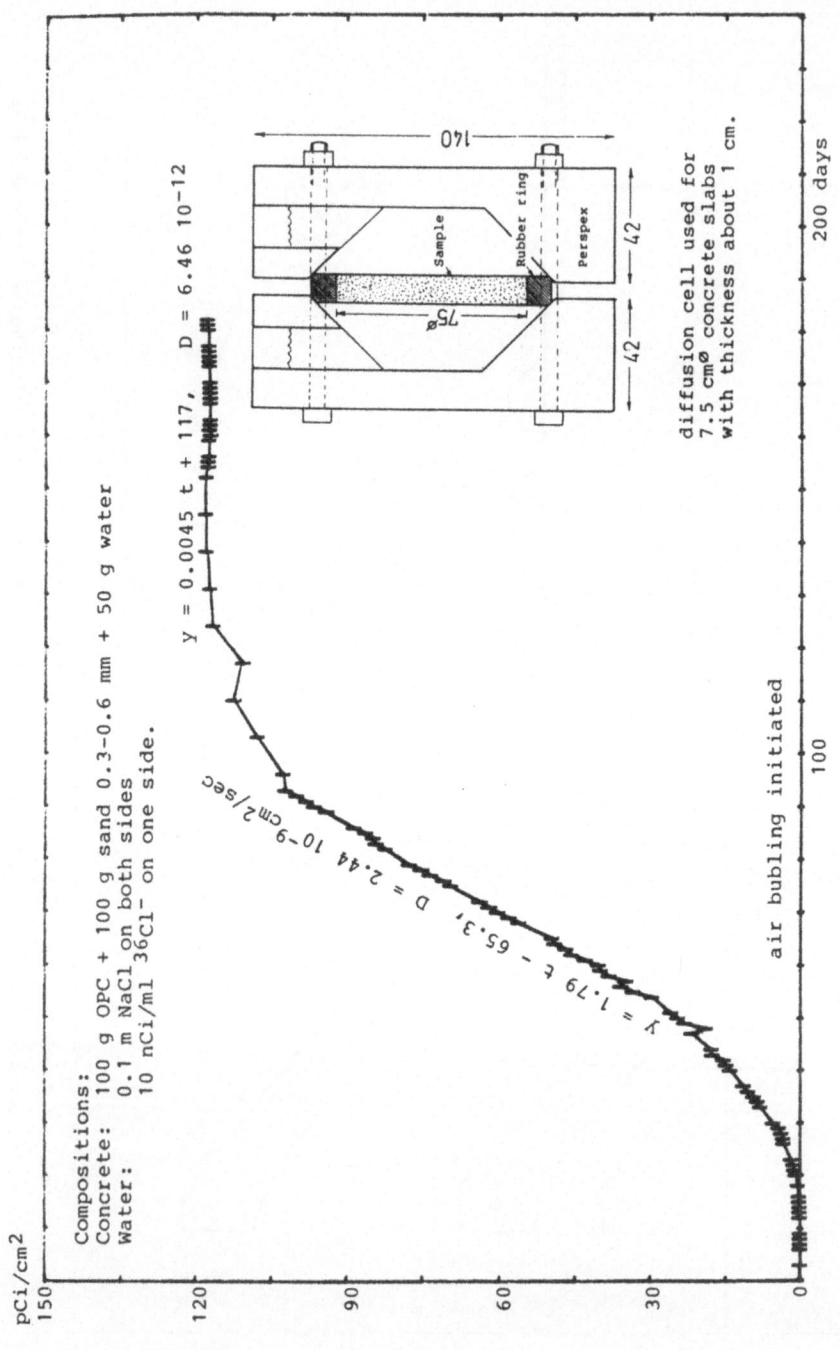

Fig. 3.16 Accumulative ^{36}Cl Activity Passed Through 1.02 cm Thick Sample of Ordinary Concrete

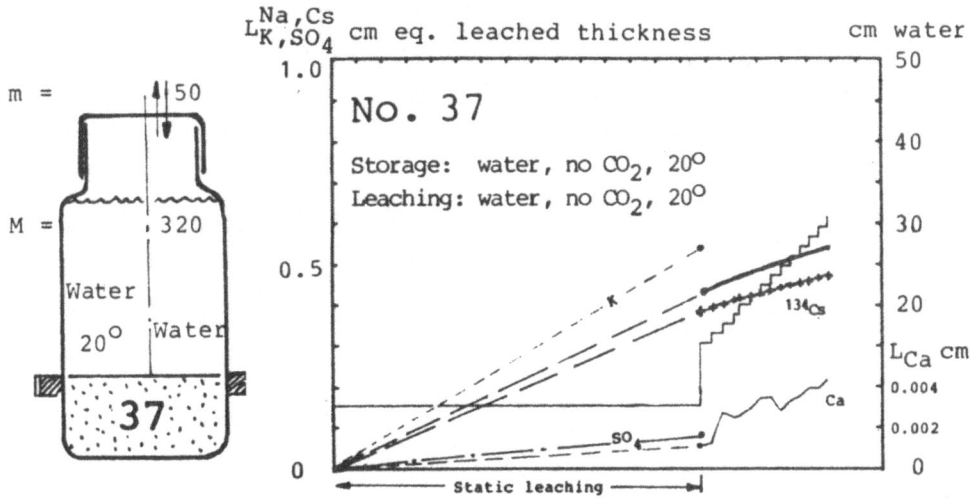

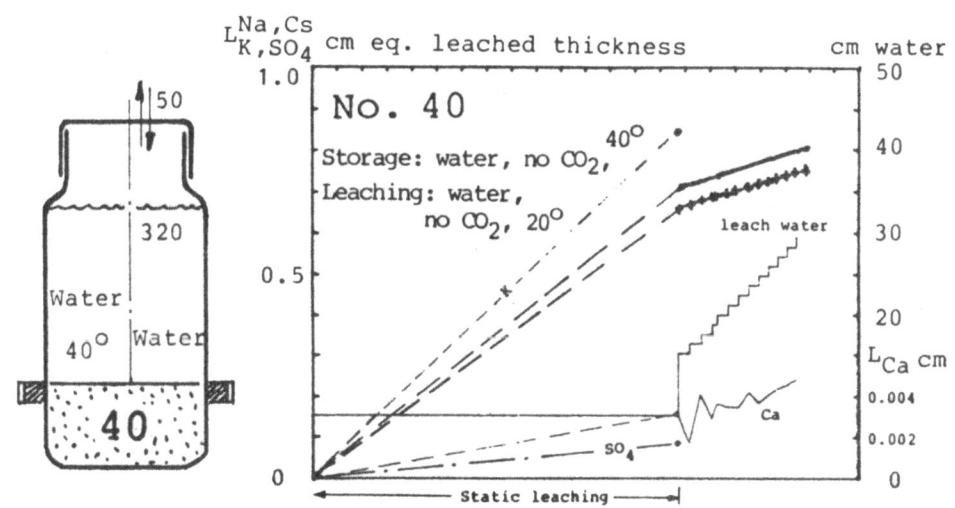

Fig. 3.17 ^{134}Cs Leaching from Sodium Sulphate Solidified in ITALIAN ORDINARY PORTLAND CEMENT

Leaching after Static Leaching in Water

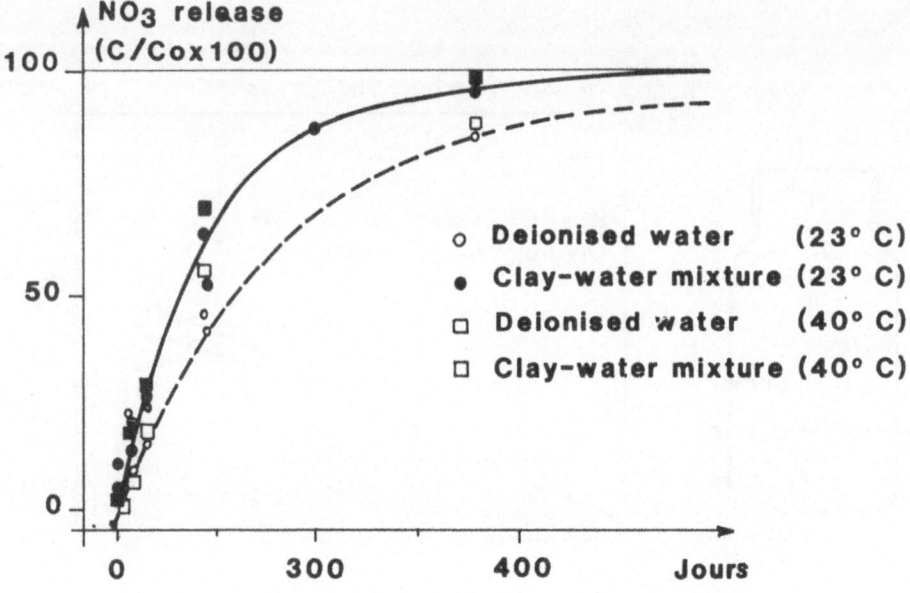

Fig. 3.18 NO3 Release from Eurobitumen Samples

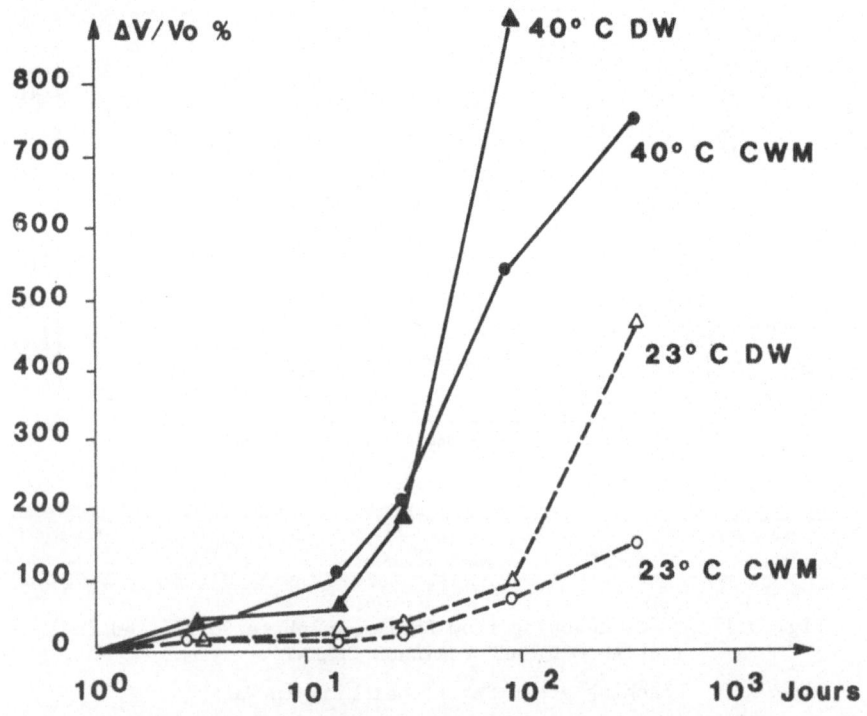

Fig. 3.19 Swelling of Eurobitumen Samples

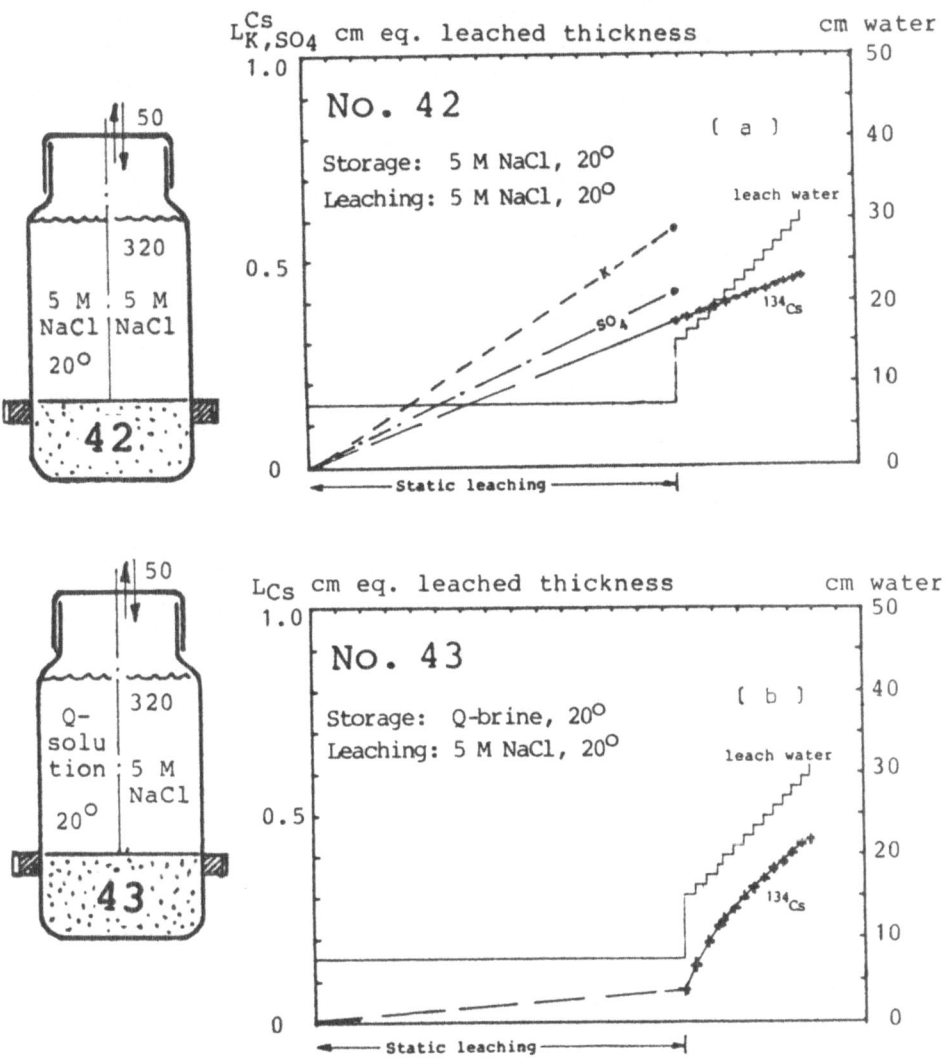

Fig. 3.20 ^{134}Cs Leaching from Sodium Sulphate Solidified in ITALIAN ORDINARY PORTLAND CEMENT

Leaching after static leaching in 5 M NaCl (a)
in Q-brine (b)

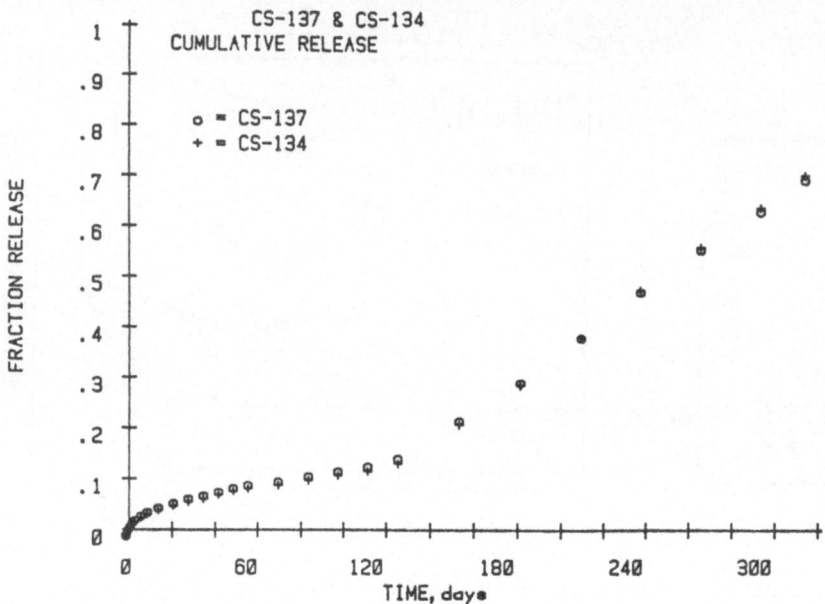

Fig. 3.21   Cumulative Fraction Release of Cs-137 and Cs-134 as a Function
of Time for BWR Evaporator Concentrate plus Ion Exchange Resin
Waste Solidified in Portland Type III Cement

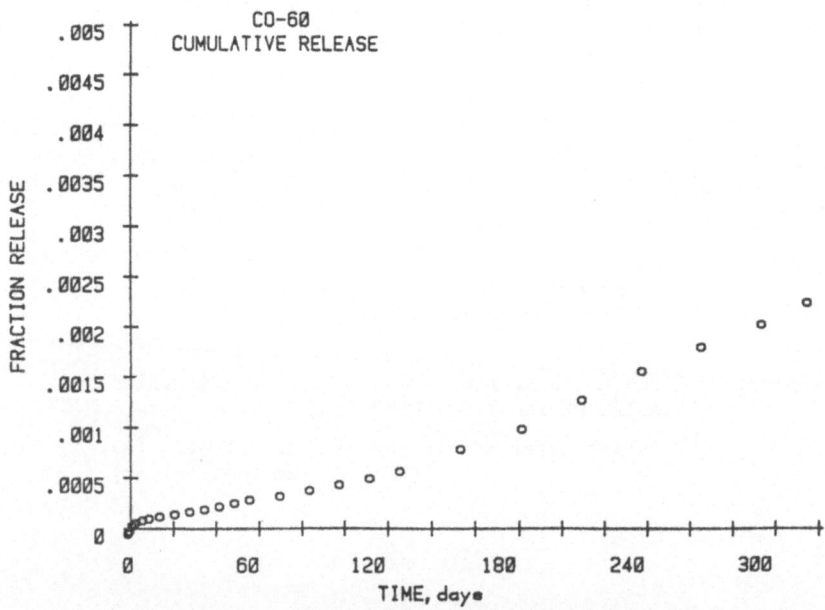

Fig. 3.22   Cumulative Fraction Release of Co-60 as a Function of Time for
BWR Evaporator Concentrate plus Ion Exchange Resin Waste
Solidified in Portland Type III Cement

SESSION 4

PROBLEMS RELATED TO NATIONAL REGULATIONS, AND
REQUIREMENTS. ASSESSMENT OF NATIONAL ACCEPTANCE CRITERIA

SYNTHESIS OF SESSION 4

Chairman : Mr L. Barrett, US Nuclear Regulatoty Commission.

Subjects of discussion : Problems related to national regulations and
requirements. Assessment of national
acceptance criteria.

Scope : Review of characterization requirements needed to define site
specific acceptance criteria.

SUMMARY OF DISCUSSION

4.1 - Problems Related To National Regulation Requirements*

4.1.1 - United States Waste Management Requirements

The U.S. Nuclear Regulatory Commission (NRC) regulates the disposal of
radioactive wastes under the Atomic Energy Act of 1954, as amended,
and the Nuclear Waste Policy Act of 1982. These acts provide the NRC
with the authority to regulate various types of radioactive waste
materials. These different categories of wastes are summarized below :

1. Wastes having activities below regulatory concern. These wastes are
regulated on a case-by-case basis.

2. Wastes requiring semi-controlled disposal. This category includes
certain disposals taking place at the waste generator's site under 10
CFR § 20.302.

3. Low-level wastes (LLW) which are suitable for shallow land burial
(Classes A, B, and C). Low-level wastes are defined in the regulation
10 CFR Part 61, "Licensing Requirements for Land Disposal of
Radioactive Waste," (Ref. $\lceil$ 4.1 $\rceil$), as radioactive wastes not
classified as high-level wastes, transuranic wastes or uranium and
thorium tailings and wastes. The maximum concentration levels for
Class A, B, and C wastes suitable for shallow land burial are defined
in 10 CFR § 61.55. Class A wastes are the lower activity wastes and
must meet the minimum waste form requirements in 10 CFR § 61.56(a) and
be segregated from stable waste. Class B wastes have higher activities
than do the Class A wastes and must meet the minimum waste form
requirements in 10 CFR § 61.56(a) and also the stability requirements
in 10 CFR 61.56(b). Class C wastes have higher activities than do the
Class B wastes and must meet the minimum waste form requirements in
10 CFR § 61.56(a) and the stability requirements in 10 CFR § 61.56(b).

---

* US NRC contribution kindly reviewed by Mr. T.C. Johnson.

In addition, disposal must also provide a barrier to inadverdent intrusion. The NRC has prepared guidance to waste generators for demonstrating compliance with the waste form stability requirements. This guidance is presented in the Technical Position of Waste Form (Ref. $\lfloor$ 4.2 $\rfloor$). Table IV.1 summarizes the tests and recommended results given in the Technical Position. An institutional control period of 100 years is provided at the low-level waste disposal site.

4. Low-level wastes with activities which exceed the 10 CFR Part 61 Class C concentrations. The rule, 10 CFR Part 61, allows for the case-by-case review of these wastes for disposal by shallow land burial.

5. Transuranic wastes (TRU) contain long-lived transuranic nuclides in concentrations greater than 100 nCi/gm. Transuranic wastes are considered to be generally unsuitable for shallow land burial.

6. High-level wastes (HLW) which are irradiated reactor fuel, liquid wastes resulting from first solvent extraction stages in a fuel reprocessing facility or solids into which the above liquide wastes are converted. Requirements for the disposal of high-level wastes are presented in 10 CFR Part 60, "Disposal of High-Level Radioactive Wastes in Geologic Repositories" (Ref. $\lfloor$ 4.3 $\rfloor$). This regulation sets performance objectives for waste package containment (300 to 1000 years), the engineered barrier system release rate (one part in 10^5), groundwater travel time (1000 years), protection against radiation exposure to workers and the public during operations, waste retrievability during the operational and performance confirmation periods, and compliance with U.S. Environmental Protection Agency environmental standards.

7. Uranium mill tailings are regulated under 10 CFR Part 40. Specific disposal requirements are given in 10 CFR Part 40, Appendix A.

Since 1983 the sources of low-level wastes disposed of at commercial disposal sites can be characterized by volume as follows :

- Class A (unstabilized) 90.6 %,
- Class A (stabilized)   6.5 %,
- Class B   2.5 %,
- Class C   0.4 %,
- Above Class C   1 %.

For light-water reactor ion exchange resins and filter sludges, many waste generators are using high integrity containers rather than solidification to provide stability for Class B and C wastes.

4.1.2 - United States Low-level Waste Leachability Requirements

The low-level waste regulation, 10 CFR Part 61, provides performance objectives for low-level waste disposal facilities. One of these performance objectives specifies that radioactive materials may not be released to the general environment to result in doses exceeding 25 mRems to the whole body, 75 mRems to the thyroid and 25 mRems to any other organ of any member of the public.

The NRC Staff has performed extensive pathways evaluations for reference disposal facilities which meet the 10 CFR Part 61 site suitability, design and operations requirements. These evaluations (Ref. $\lceil$ 4.4 $\rceil$, $\lceil$ 4.5 $\rceil$) show that groundwater pathways are not limiting. Intruder scenarios become the basis for the radionuclide concentration values in the waste classification system.

Since the groundwater pathways are not limiting for a 10 CFR Part 61 disposal site, it was unnecessary to assign highly restrictive leach requirements on wastes. The NRC Staff studies, however, did show that waste products which had very poor leach resistance should be avoided. Our sensitivity studies verified that a leach index of 6, as calculated by the method presented in the ANS 16.1 standard (Ref. $\lceil$ 4.6 $\rceil$) would be acceptable. These studies are, therefore, the basis for selecting this value in the Technical Position on Waste Form.

4.1.3 - United Kingdom National Regulation Requirements

In the United Kingdom, the handling of radioactive material is governed by the Radioactive Substances Act 1960 and additions.

The UK Department of the Environment has issued a draft report $\lceil$ 4.7 $\rceil$; a copy of which was distributed to the workshop participants. This document gives the views of the Authorizing Departments under the Radioactive Substances Act 1960 regarding the principles to follow in assessing proposals for land disposal facilities (Low and Intermediate Level Wastes). The different waste categories are defined and their disposal is regulated as follows :

- De minimis Level wastes, with an activity of less than 0.1 millicurie per metre cubed are disposed of with normal local authorithy solid waste.

In addition incineration of waste material in up to millicuries amounts is permitted, level dependent on circumstances.

Larger disposals of solid waste in millicurie amounts is permitted by burial under special precautions. These burials are negotiated individually and occur at a large number of sites.

- Low-level liquid wastes, low-level liquid wastes containing up to millicurie amounts of beta gamma emitters are, subject to individually negotiated conditions, often disposed of directly to the sewerage system".

- The DRIGG shallow land burial site, the DRIGG shallow land burial site is permitted to accept solid wastes with an alpha activity of up to 20 millicuries per metre cubed and up to 60 millicuries per metre cubed of beta emitting material, subject to the further condition that the surface dose rate for beta gamma emitting material must not exceed 0.75 rads per hour.

Up to 1.0 $Ci.t^{-1}$ were until recently dumped into the sea under the London Convention (OCDE-AEN).

Two land burial sites, ELSTOW (BEDFORDSHIRE) and BILLINGHAM* are under discussion for MLW disposal. NIREX aims for a target of a maximum admissible operator dose rate of 100 mRem.year^{-1} for ELSTOW and 10 mRem.year^{-1} for BILLINGHAM. The general Policy is to postpone the definition of waste form specifications for shallow land burial until practical in situ experience has been gained. The difficulty at the moment is to define minimum requirements for waste conditioning before the disposal site is known.

As there is no land disposal facility for MLW in the UK, the Regulatory Bodies are collecting and evaluating proposals for extended interim storage. From these proposals preliminary requirements for the waste packages can be deduced ; according to a study on the radiological impact of a storage site on the environment performed by NRPB. The major short term risk is disruption due to human intrusion some hundreds of years after closure. For this reason, a site control is anticipated for the same duration. The use of high integrity containers is still under discussion (major problem is gas generation). This question affects the ranking of characterization requirements (leach resistance, gas permeability). The multibarrier concept for control of activity release is recommended. The provisions for waste disposal will most probably include the following waste characteristics :

- High integrity.
- Low specific surface area.
- Chemical and biological stability.
- Low leachability.
- Radiation stability.
- Absence of critical hazards.
- Specific gravity of waste package > 1.2 (related to see dumping).

4.1.4 - SWITZERLAND national requirements

Solid radioactive wastes are defined in general to have activities above the following limits :

	Concentrations (pCi.g^{-1})	Contamination (pCi.cm^{-2})
beta/gamma activity :	20.000	100
alpha activity :	20	10

---

* The BILLIGHAM site has now been ruled out as a possibility for a repository.

These values are derived from a general dose limit of 500 mrem.y^{-1}.

For gaseous and liquid effluents, concentration limits are also defined based on a dose of 100 mrem.y^{-1}.

In order to limit further the doses arising from waste production on a larger scale in the nuclear field various additional limits have been set ; for instance for nuclear power plant wastes, 50 µrem.h^{-1} at 10 cm, and also specific activities lower than 2.000 pCi.g^{-1} beta/gamma if it can not be proven that the associated dose will stay under 10 mrem.y^{-1}. For these nuclear plants the gaseous and liquid effluents must not lead to doses greater than 20 mrem.y^{-1}.

Treatment, solidification and packages of the wastes are subject to regulatory body directives and controls. The leach rate over 150 days must be smaller than 10^{-3} g.cm^{-2}.day^{-1}.

Final disposal of the waste in underground repositories after closure must not lead to individual dosis greater than 10 mrem.y^{-1} at any time.

According to the disposal projects of Nagra in which both technical and geological barriers play important roles, the requirements on the waste packages therefore will be rather limited in scope and will mainly be on the activity source (i.e. concentration limits).

4.1.5 - <u>SWEDEN National Position</u>

The purpose of disposal of radioactive waste into shallow ground or rock cavities is, as everywhere, to limit the risk for an unacceptable release of the radionuclides in the accessible environment. The multibarrier system is taken into consideration, with two main aspects : physical and chemical barrier. The physical barrier effect is to limit or hinder the water flow through the barriers by using materiel having a low or zero permeability (examples : clay-buffer for low permeability ; closed metal container for zero permeability). The chemical barrier is obtained by stability limitations, ion exchange and sorption.

<u>In SWEDEN, the major concern for quality assurance</u> in waste disposal in the long term function of the engineered barriers which is essentially dependant of the swelling, of the gas generation of the waste forms (reactors and decommissioning waste) and possible external chemical attack. In order to obtain a thorough knowledge of the quality and the conditions of the confinement when the repository is being closed, it is aimed for a complete picture claiming for :

- General description of the management scheme concerning : Treatment - Handling - Interim storage - Transport - Disposal.

- Safety analysis and derivation of functional requirements : Mechanical stability - Environmental conditions - Analytical conditions - Radiological properties.

- Choice of operational requirements (based on functional needs) in view of : Waste package (cement container in most cases) - Waste form - Radioisotope inventory.

- Accomplishment of the realisation : Plant description - Operational process - Package specification - Waste material characteristics - R and D results.

- Verification of functional requirements : Product quality control (sampling) - Compliance with imposed limits. About this quality control, the properties to be checked by operators are :

. the container, regarding its water tightness, its frost resistance and the compression strength of the material,

. the waste form, in view to the homogeneity, density, compression strength and activity.

In general, the Swedish Authorities are much more concerned with the structural durability of the waste package than with the leach resistance of the waste form.

4.1.6 - Federal Republic of Germany (FRG)

In the FRG two repository sites are under investigation :

a) the Konrad mine (old iron ore mine),
b) the Gorleben salt dome.

The Asse salt mine that has been used for the disposal of radioactive wastes until the timely limited license expired at the end of 1978 is now used as a R+D facility to provide data relevant for licensing the planned repository in Gorleben. The future use of the Asse salt mine will be discussed again in 1987.

The planning of the repositories and the safety assessments are based on the safety criteria for the disposal of radioactive wastes in mined repositories published by the Federal Minister of the Interior.

These criteria state e.g. :

- the safety of a repository has to be proven by a site specific safety analysis,

- the limiting values of the Radiation Protection Ordinance shall not be exceeded in the normal operation of the repository and during incidents in the operational phase,

- in the post-operational phase the radiation exposure shall not exceed the values of the 30 mrem/a-concept.

a) The Konrad mine

The Konrad iron ore mine has been investigated by the GSF
(Gesellschaft für Strahlen- und Umweltforschung) until mid 1982. The
PTB (Physikalisch-Technische Bundesanstalt) has applied to license
this site for the disposal of radioactive wastes in August 1982 at the
responsible licensing authority. The completition of the application
documents is foreseen for the end of 1985. Operation of the repository
is expected for the year 1989.

The status of the work can be summarized as follows :

α) safety assessments :

- the radioactive wastes that are arising or are to be expected in the
future have been summarized including their relevant data,

- a site specific safety assessment has been carried out with the data
of the radioactive wastes for the operational phase of the repository
(normal operation and incidents),

- the main results of the safety assessment were, that the various
radioactive wastes categories can be condensed into six waste product
groups and two waste classes. Limiting values of the radionuclide
contents in the waste packages have been derived. All the radioactive
wastes without significant heat production are suitable for the
disposal in the Konrad iron ore mine,

- the post-operational phase is intended to be analyzed until the end
of 1985. A deep drilling has been necessary to provide relevant
informations on the geological and hydrogeological situation. The
possibility of a radionuclide release has to be analyzed. If it cannot
be precluded, possible consequences of a release have to be
calculated.

β) Product control

- a compact concept for the quality control of radioactive wastes has
been developed. The most important features are the necessity of
non-destructive or destructive test have to be applied to existing
waste packages. Waste package quality control should be performed by
the inspection of the conditionning processes that have been qualified
and well instrumented,

- 14 relevant properties for the quality control have been identified.
Not all properties apply to each waste package. The leach/corrosion
rate are excluded because their relevance can only be identified after
the completion of the post-operational safety assessment,

- a quality control group which is active on behalf of the PTB has
been established in the KFA (Kernforschungsanlage), Jülich.

b) The Gorleben salt dome

The Gorleben salt dome has been selected for investigating its suitability for the disposal of all types of radioactive wastes. The investigation from above ground e.g. by drilling and seismic measures has been terminated. Suitable points for the sinking of the shafts have been identified.

Shaft sinking has begun after a positive vote for the continuation of the exploratory work. The drilling of the freezing holes for the shaft sinking in nearly completed. The underground investigation is planned to be terminated in about 1990. The operation is scheduled about the year 2000.

Safety assessments on the basis of model assumptions for the repository have been carried out and published $\underline{/}$ 4.8, 4.9 $\underline{/}$. They include the operational phase of the repository and the stability calculations for the geological situation due to thermal and thermomechanical effects from the heat generating wastes as well as unlikely geological events.

A scenario for a water intrusion into the repository which cannot be precluded due to the effects from the heat production of radioactive wastes has been established. The development of a computer programme for the modelling of the radionuclide releases from repositories has been performed in parallel. This programme has been applied for test purposes to the planned repository in Gorleben. The results of the above-ground exploration, including the site specific sorption data of radionuclides, and model assumptions for the repository and the underground situation have been used. The modelling work has been performed within the Projekt Sicherheitsstudien Entsorgung (PSE) on behalf of the BMFT (Bundesminister für Forschung und Technologie). The main results of these preliminary calculations are :

- the expected doses from possibly occuring radionuclide releases are below the limiting values,

- the most relevant radionuclides are Tc-99 and Np-237,

- the maxima of these releases accur before 20.000 years after the closure of the repository (see fig. 4.1),

- the high-level wastes do not contribute to the radionuclide releases because the high-level waste bore-holes are closed at the given repository layout when the intruding brines reach the high-level waste disposal area.

c) Conditions for leach and corrosion experiments

The conditions for leach/corrosion experiments have been discussed in a group of experts. In a first attempt 11 brines were evaluated and classified into two priority groups which cover all possibly occuring compositions in the northerm German salt domes [4.9]. In a second attempt the result of experiments performed were analyzed with the intention to deminish the number of brines for experiments. A general recommendation could not be given due the different state of experimental investigations of the different materials (e.g. glass, cement/concrete, metals and spent fuel).

The composition of the occuring deep waters in the iron ore mine Konrad has been analyzed by the GSF and is applied to leach/corrosion experiments.

4.1.7 - Japanese situation

The maximum permissible dose rate for the public from a LWR reactor site in JAPAN is practically maintained to 5 mRem.year^{-1}. At the time being, no site has yet been licensed for LLW and MLW disposal and no requirements have been determined and set forth for MLW conditioning up to now. Utilities (reactor sites...) who want to condition their wastes, mainly LLW, have specified preliminary checks for minimum requirements for conditioned cement and bitumen wastes which relate at least to :

- Homogeneity.
- Water content.
- Compressive strength.
- Surface dose.
- Flammability.

In general, the utilities are looking for the largest possible volume reduction and the lowest possible release rates.

4.1.8 - ITALY position for waste packages control and disposal

No disposal sites have until now been selected in ITALY for MLW disposal. The waste packages are presently stored at the nuclear facilities sites (mainly electronuclear power stations and nuclear research centers). In the past, ITALY partecipated to a sea dumping operation, and the sea dumping could be till now taken into consideration for the MLW disposal. From this point of view, the conditioned waste shall meet the requirements established by NEA for such a type of operations.

At present, a guideline for the waste conditioning is in preparation at the ENEA-DISP, that is the italian Regulatory Body. The following conditioned waste characteristics will be taken into considerations :

- minimum compressive strength,
- thermal cycling resistance,
- radiation resistance,
- fire resistance,
- leaching resistance,
- biodegradation resistance,
- immersion resistance.

Such a document will be probably issued in the next months.

4.1.9 - Policy, and Regulatory aspects·in FRANCE

The regulatory body in FRANCE is the Service Central de Sûreté des Installations Nucléaires (SCSIN), operating inside the Ministère du Redéploiement Industriel et du Commerce Extérieur (Ministry of Industry and International Trade). The SCSIN relies on the IPSN (Institut de Protection et de Sûreté Nucléaire) of the CEA for technical advices.

SCSIN defines and edicts the "Règles Fondamentales de Sûreté" (RFS : Fondamental Safety Rules) and for designated disposal sites, the "Prescriptions Techniques" (Technical Prescriptions) ; the site operators, within the limits of the RFS must respect the Technical Prescriptions.

ANDRA, the French Agency, for Radwaste Management, is responsible, inside the CEA, of the Radwaste disposal for operation time and for the long term. In charge of the disposal sites (only one, Centre Manche, a shallow land burial site is in operation), ANDRA defines the Specifications Techniques for agreement of waste package before disposal.

The safety assessment of long term shallow land burial disposal sites is mainly based on the radionuclides return to biosphere through two ways : atmospheric dust in case of intrusion and surface waters through leaching of waste and flowing through barriers and soils, taking into account adsorption and retardation properties of these barriers and soils.

For this second aspect (leaching and contamination of water surfaces), the safety requirements involve that :

a) In the disposal site specific risk assessment, two groups of parameters are taken into account :

- site parameters : permeability of the geological environment, retention properties, performances of engineered barriers, local pluviometry, hydrology, etc...,

- waste form characteristics : radionuclides inventory, leaching and ageing properties, container effect, reaction in the waste form and waste form with barriers or soil, irradiation long term effects ... etc. It is of special importance to know the time dependance behaviour laws of these characteristics.

b) The waste characteristics remain intrinsically better than the imposed limits. This requirement is due to the fact that :

- various uncertainties (on the model, on the site parameters, on the waste behaviour) affect the risk assessment. So it is necessary for the barriers to have minimal independent performances,

- the conditioning matrices should have been designed and applied to waste independently from the predefined conditions.

From the preceeding considerations, it results that :

a) The leaching tests must be performed in a laboratory following a standard procedure ; the test results will be what we call the intrinsic leaching characteristics.

b) The leaching mechanisms, have to be understood and modelled so that the laboratory results could be extrapolated to the different site conditions in order to allow the risk evaluation. So far, it is excluded, because non realistic, that systematic tests be performed on each site. Meanwhile some in situ tests, like lysimetry tests for example, can be done in order to help the phenomena understanding and the establishment of a "transfer function" from the laboratory leaching rate and the in situ leaching rate.

c) The time-dependant leaching rate mechanisms (including the matrix degradation, the chemical environment evolution,...) must be understood in order to improve the disposal risk evaluations.

Fundamental Safety Rules (RFS)

Shallow land Burial

- RFS I.2 Rev. 1 has been published June 19, 1984. "Bases de conception pour les centres de surface".

Main features : Wastes concerned, multibarrier concept, post closure duration survey (300 years), limitation of α specific activity : average for the site < 0.01 $Ci.t^{-1}$ with a normal limit < 0.1 $Ci.t^{-1}$ for one package and never above > 0.5 $Ci.t^{-1}$, control of transfers to waters, to atmosphere.

- RFS I.2 b - Serie "U" Not yet published but probably applicable mid-85. "Conditions préalables à l'agrément des déchets solides destinés à être stockés en surface".

Main features : Wastes concerned (< 0.1 Ci $\alpha.t^{-1}$), waste packages : homogeneous, heterogeneous, required characteristics :

. Leaching resistance (homogeneous waste forms)
< 10 $Ci.t^{-1}$ $\beta\gamma$: leach rate $< 5.10^{-4}$ $cm.d^{-1}$
> 10 $Ci.t^{-1}$ $\beta\gamma$: leach rate $< 2,5.10^{-4}$ $cm.d^{-1}$
< 0.01 $Ci.t^{-1}$ α : leach rate $< 5.10^{-6}$ $cm.d^{-1}$

. Under a pressure of 0,35 MPa, the deformation of the package must not exceed 3 %.

. Thermal cycling resistance (- 20° + 5° - and + 5° - 40 ° in presence of water) 2 times 5 cycles.

. Radiation resistance : at least 1.10^7 rads with a dose rate $< 2.10^4$ rad.h^{-1}.

The RFS should be reviewed five years after publication in order to take account the progress in the understanding of basic phenomena.

- <u>RFS concerning the reprocessing waste forms</u>

. <u>RFS III.2.a</u> : General conditions - Published : Sept. 24, 1982. Contain a waste classification for long terme disposal.

Category I   : < 0.1 Ci α.t^{-1}
Category II   : 0.1 to 1 Ci α.t^{-1}
Category III : > 1.0 Ci α.t^{-1}

(inside this category, vitrified HLW are distinguished).

. <u>RFS III.2.b</u> : Dispositions applicable to HLW glass - Published Nov. 12, 1982.

. <u>RFS III.2.c</u> : Dispositions applicable to bituminized waste forms - Published April 4, 1984.

. <u>RFS III.2.D</u> : Dispositions applicable to cemented waste forms - Published Feb. 1, 1985.

- <u>Technical specifications established by ANDRA</u>

Based on Technical Prescription from SCSIN and respecting the RFS imposed contraints and limits, they are specific of a disposal site. For the only existing disposal site of Centre Manche, two specifications have been published.

- <u>Waste packages resulting from cementation of waste inside a reinforced concrete container</u> (giving mechanical resistance, leach rates for Cs, Sr, Co, α nuclides, shock and fall resistance container compressive strength and permeability).

- <u>Waste package resulting from embedding of ion exchange resins in a thermosetting polymer matrix in a steel drum</u> (giving mechanical resistance filling degree, leach rates for Cs, Sr, Co thermal cycling resistance, shock and fall resistance, irradiation resistance and waste package maximum dimensions. Standardized control procedure for leach rate and thermal cycling are also given).

4.1.10 - Situation in CANADA (A.S. WILLIAMSON)

"In CANADA the responsibility for radioactive waste management is divided between the Federal and Provincial governments. Spent fuel disposal is a Federal matter and Atomic Energy of CANADA has responsibility for development of fuel conditioning treatments and selection and operation of a fuel waste disposal facility. The Ontario government and Ontario Hydro have responsibility for conditioning, storage and eventual disposal of all other power reactor wastes. At present, Ontario Hydro practices conditioning and intermediate storage of LLW and MLW. Ontario Hydro is investigating the concept of disposal of these wastes in concrete trenches in shallow land burial or rock caverns at intermediate depth, however formal proposals to operate a facility have not yet been presented to regulatory and licensing agencies. Additionally regulations for the disposal of any radioactive waste, disposal facility operating requirements and permissible release limits from repositories have yet to be established and specified".

4.1.11 - Situation in BELGIUM (J. CLAES)

Until 1982, Belgium has evacuated its LLW and partly ILW by sea disposal, in the frame of the London dumping convention and the NEA multilateral mechanism.

Today however, this disposal route has become uncertain for other than scientific reasons and management programmes are being considered for land disposal of these wastes, including intermediate storage to allow for the preparation of a land based alternative.

When processing the LLW and ILW for sea disposal, Belgium applies at present the best available techniques for volume reduction and conditioning. Therefore, it is believed that there are no fundamental reasons to modify substantially the present conditioning procedures, although additional efforts on further volume reduction are done.

Since the number of waste packages and waste forms are relatively important, NIRAS/ONDRAF is now limiting its waste acceptance criteria to quality control of the outer containers (galvanised metallic 400 L-DRUMS as reference container) at their manufacturing and inspection of the waste packages at the waste conditioning plant, based upon the present conditioning practices. Reinforcement of these criteria will eventually be decided upon, when the safety assessments on a land disposal facility will have been performed.

4.2 - Main points of interest in the discussion

4.2.1 - Importance of leaching resistance evaluation for long term disposal

Several experts have pointed out that in their country, leaching resistance is not the most important criterium : USA, SWITZERLAND ; for others, it presents the same importance as the other criteria : FRG. But, however, where regulatory dispositions for surface disposal have been published, (USA, FRANCE) leaching resistance has been given figures of merit : leachability index, maximum acceptable leach rates.

In fact, two main concepts for safety assessments of closed repositories are taken into account : man inadvertent or voluntary intrusion, giving rise to dust contamination of atmosphere and water intrusion, causing leaching and transfer of radionuclides as a "source term" through barriers and ground layers to surface waters.

Whatever is the national status of regulatory aspects, it seems that leaching resistance must be evaluated in the disposal site conditions and also, for the handling transport and operating risks as accidental eventuality.

4.2.2 - Role of barriers

It has been said that they may be multiple (several experts) ; that they have to play their role independently (FRANCE), that they can play two kinds of roles : a preventive by low or zero permeability and a remedial by chemical or physico-chemical exchanges or sorption (SWEDEN). The characterization of barriers, specially in the case where these constitute the only safety system, is therefore very important.

4.2.3 - Possible strategy for selection of a methodology

The approaches presented during Session 3 may lead us to the following reflexions.

What is the most suitable methodology to acquire reliable data for the safety assessment of disposal ?

NAGRA (SWITZERLAND) is employing for HLW forms a methodology which might be adaptable to MLW forms.

- Start up with one waste form only.

- Perform leach testing with active specimens in distilled water (standard method) and make a comparative testing of inactive specimens. If reasonably comparable, continue for model preparation and validation, using mainly inactive specimens.

- Define the expected disposal conditions.

- Check wether the release mechanisms can be modeled under repository conditions.

- Develop a model for long term prediction and run validation tests under simulated disposal conditions. The modeling objectives are to explain the fractional releases resulting from the source term determined (not forgetting the nature of releases radionuclides : speciation).

The modeling must also take into account the relevant conditions : short term test in view of extrapolation, suitable S/V ratios, use of parameters based on natural analogues ; measurement required and validity evaluation of basic mechanisms. Such a methodology can help to :

- The effects of the waste on the transport properties through the other barriers must be established. Three factors are important :

. the chemical form of the released radionuclides,
. the release of nonradioactive material which could have a chemical influence on the other barriers.

In the establishment of the chemical form of the released radionuclides and of the release of nonradioactive material due consideration must be given to the effects of the ageing, e.g. due to radiolysis or microbial attack. This is particularly true for the long-lived alpha-containing waste.

. Changes in the waste packages that could mechanically disturb the other barriers, e.g. gas production, swelling or collapse.

- Close the gap between experiment and model.

- Obtain agreement in the interpretation between experimentalists and modelers, and further regulatory and implementary bodies.

This methodology, after discussion, has seemed a little optimistic to some experts when applied to MLW, but promising if it could be tested in several countries with exchange of information.

4.2.4 - <u>Possible conclusions for session 4</u>
         (prepared after the workshop)

From a summary made of a paper from SWEDEN, we could draw the following conclusions about what are the important results to be obtained from waste characterization tests with regard to safety assessment :

- The transport properties of the radionuclides in the waste under repository conditions must be established. These consist of two parts :

. the solubility of the radioactive material in the water getting in through the matrix ; the composition of this water could change with time and this change must be established,

. the transport properties of species through the matrix, including both physical and chemical effects ; these will also change with time which has to be estimated.

With this information, the release of radionuclides from the waste packages (the source term) could be modeled and calculated. In this model also, data for the barrier effect of the container should be included.

REFERENCES

/ 4.1 /   U. S. Nuclear Regulatory Commission, Title 10 Code of Federal Regulations Part 61, "Licensing Requirements for Land Disposal of Radioactive Wastes", December 27, 1982.

/ 4.2 /   U. S. Nuclear Regulatory Commission, "Final Waste Classification and Waste Form Technical Position Papers", May 11, 1983.

/ 4.3 /   U. S. Nuclear Regulatory Commission, Title 10 Code of Federal Regulations Part 60, "Disposal of High-Level Radioactive Wastes in Geologic Repositories", February 25, 1981 and June 21, 1983.

/ 4.4 /   U. S. Nuclear Regulatory Commission, "Draft Environmental Impact Statement on 10 CFR Part 61 - Licensing Requirements for Land Disposal of Radioactive Waste", NUREG-0782, September 1981.

/ 4.5 /   O. I. Oztunali, et al., "Influence of Leach Rate and Other Parameters on Groundwater Migration", NUREG/CR-3130, February 1983.

/ 4.6 /   American Nuclear Society Standard, "Measurement of the Leachability of Solidified Low-Level Radioactive Wastes", Working Group ANS-16.1, June 20, 1984.

/ 4.7 /   UK.DOE.
          "Disposal Facilities on Land For Low and Intermediate Level Wastes - Draft Principles For the Protection of the Human Environment".
          UK-DOE 1983.

/ 4.8 /   E. Warnecke, H. Illi, Evaluation of Product Specifications
           with a Safety Analysis for a Disposal Mine, in : J. G Moore
           (ed.), Scientific Basis for Nuclear Waste Management, Vol.
           3, p. 19-26, Plenum Press, New York/London (1981).

/ 4.9 /   E. Warnecke, H. Illi, D. Ehrlich, Requirements for the
           Disposal of High-Level Radioactive Wastes, in : R. Odoj,
           E. Merz (eds.), Proceedings of the International Seminar on
           Chemistry and Process Engineering for High-Level Liquid
           Waste Solidification, Jülich, June 1-5, 1981, Berichte der
           Kernforschungsanlage Jülich Jül-Conf-42 (Vol. 2),
           p. 892-815, Jülich (1981).

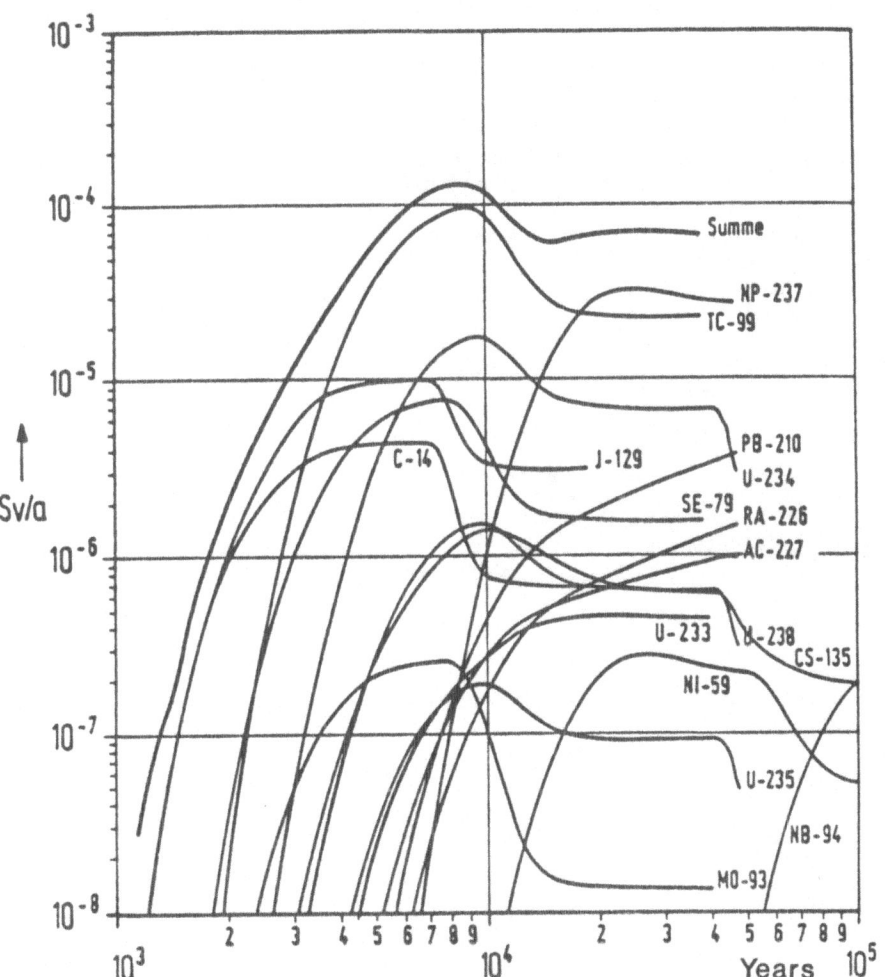

Fig. 4.1   Dose Rates of Various Radio-Nuclides Versus Time and
            Contribution to the Total Dose Rate

TABLE IV.1
WASTE FORM STABILITY CRITERIA

Tests methods and minimum parameters values

Criteria	Test methods	Parameters values
- Free standing monolith	-	-
- Free liquid	ANS 55	H_2O < 0.5 % by volume 4 < pH < 11
- Compressive strength	ASTM C.39 or ASTM D.1074 (bitumen)	⩾ 50 PSI
- Irradiation resistance (compressive strength after 1.10^8 Rads γ)	Same methods	⩾ 50 PSI
- Biodegradation (Bacteria - Fungus)	ASTM G.21, ASTM G.22	No culture growth visible, if yes additionnal testing
. Compressive strength	ASTM C.39/ASTM D.1074	⩾ 50 PSI
- Leach testing	ANS 16.1 minimum 90 days	Leachability index ⩾ 6
- 90 days immersion test		
. Compressive strength	ASTM C.39/ASTM D.1074	⩾ 50 PSI
- Thermal Cycling		
. Compressive strength	Same methods	⩾ 50 PSI
- Homogeneity . Compressive strength (CS)	Destructive analysis of full scale specimen	CS ⩾ 50 PSI for every regions

SESSION 5

PROPOSALS AND RECOMMENDATIONS FOR FUTURE ACTIVITIES
WITH RESPECT TO THE DEFINITION OF LEACHING MECHANISMS
(INTERNATIONAL COOPERATION)

SYNTHESIS OF SESSION 5

Chairman: Dr. R. Köster, KFK Karlsruhe

Subject of discussion: Proposals and recommendations for future
activities with respect to the definition of leaching mechanisms
(international cooperation).

Objective: Prediction of long-term behaviour for safety assessment.

Scope: Study of laboratory and in-situ, bench and full-scale, inactive
and active test relations and establishment of source-term
models.

SUMMARY OF DISCUSSION

5.1. Bench or full-scale leaching

5.1.1 Scale effects

Saclay has launched a series of experiments investigating the effects of
scale. Specimens of 0.2, 2, 20 and 200 l volume are being leach tested
under static conditions. The leachant volume is kept small, little or
no water removal is applied and the surface area to volume ratio is
being held constant for the different sizes. The first active tests
($40 \propto Ci/m^3$, $8 \beta/\delta^2$ Ci/m^3) are scheduled for 1985. Scale effects, if any,
might be due, e.g. with cements, to the difference in hydration tempera-
tures and thermal gradients arising during the curing stage. This might
have an influence on the overall porosity, pore size distribution and
incipient cracking of the material.

Investigation of scale effects has also been carried out at KFK Karlsruhe. OPC specimens with volumes of 2 and 200 l containing inactive
(simulated) research centre waste have been tested. The hydration
temperature amounting to 80-90°C for the 200 l specimens appeared to be
approximately 20°C lower for the small specimens. With the rather
restrictive number of test specimens of four per size, no inhomogeneities, cracks or difference in leach rate were observed.

Similar tests were performed at Oak Ridge with inactive specimens (volume: 0.01 and 200 l). Here too, no differences in leach rate have been
measured.

Furthermore AEE Winfrith investigated scale effects on the hydration
heat evolution of cement. Three different types of cements were tested:
OPC, ground granulated (quenched) blast furnace slag (BFS) and pulveriz-
ed fuel ash (PFA). The sizes varied from 30 to 500 l. The centre-line
temperature for all three cements was largely dependent on the size (see
Fig. 5.1).

With regard to possible inhomogeneities it was pointed out that duplica-
tion of each test is strongly recommended.

5.1.2 Water/cement ratio

Winfrith also investigated the effect of the cement replacement ratio on
the hydration heat evolution with the above mentioned cement materials

BFS and PFA. The peak temperature dependence is given in Figs. 5.2 and 5.3 respectively. The different behaviour of the two materials is shown in Fig. 5.4. This is of major importance as the leaching rate seems to depend on the hydration temperature. It was pointed out in this context that the water/cement ratio has to be kept under strict control, as it influences the porosity of the material.

5.2. Long-term radionuclide behaviour

5.2.1 Release mechanisms

In the Federal Republic of Germany the reference matrix material for MLW is considered to be OPC, the (α -waste) repository a salt mine. The rock salt contains approximately 0.07% of water. For the accidental scenario an attack of NaCl-solution and Q-brine consisting of 25 wt.% of Mg Cl_2 and approximately 2 wt.% of Mg SO_4 is assumed. Presumed interactions are:

. Leaching of OH^-, K^+, Na^+, Ca^{2+}, Cl^- and Cs^+

. Exchange reaction of
$$Mg\ Cl_2 + Ca\ (OH)_2 \rightleftharpoons Mg\ (OH)_2 + CaCl_2$$

. Volume changes due to e.g. the reaction of

$$Ca\ (OH)_2 + SO_4^{2-} \longrightarrow Ca\ SO_4 \cdot 1/2\ H_2O$$

$$or.\ 2\ H_2O$$

Some elements, like Cs, are presumed, at least initially, to be released according to the diffusion law. Others follow a dissolution controlled mechanism. The precipitation rate of Ca $(OH)_2$ in the pore system is being determined and the solubility for Ca $(OH)_2$ and $CaSO_4$ versus time are being calculated based on the exchange reaction Ca/Mg.

The German approach for leaching/corrosion experiments is, however, based on normal operation and the long-term aspects (see § 4.16).

For the aqueous pathway radionuclide release from cemented waste forms in salt brines have been investigated taking into account three corrosion processes:

- leaching of soluble compounds
- exchange reactions (Ca$(OH)_2$ + MgCl$_2 \rightleftharpoons$ CaCl$_2$ + Mg$(OH)_2$)
- formation of new phases causing swelling.

Leach experiments, X-ray analysis of phase composition and determination of element concentration profiles by electron microprobe analysis in the cement stone sample as a function of time and distance from phase boundary were carried out. These data have been integrated into a theoretical model using finite difference techniques to establish an improved basis for long-term predictions. A comparison between measured and calculated element concentration profiles is shown in Fig. 5.5.

The Brookhaven National Laboratory started a DOE sponsored programme on determining the leaching mechanisms of a variety of matrix materials in use or being considered for the immobilization of low and medium level waste. The purpose of the exercise is to investigate the prevailing leaching mechanisms in order to obtain a basis for predicting the long-term behaviour in a disposal environment. This will provide guidance for the use and choice of immobilization materials, the treatment of the different waste streams, the optimal waste/matrix ratio and the suitability of waste form containers. The goals are to study the behaviour of the materials properties, to identify controlling properties which influence leaching mechanisms, to determine and validate appropriate mathematical models and to establish a methodology for predicting leaching.

First a literature survey will be carried out concentrating on:

- Screening and compilation of published leaching data for inclusion in a data base.

- Identification of existing mass transport leaching models (2).

- Evaluation of models.

In parallel, leaching experiments on selected waste forms will be performed to:

. Provide long-term baseline information on the pure matrix materials

. Resolve dominant long-term trends

. Establish a measure for the overall statistical variability of leaching experiments.

The selected matrix materials are:

- Portland cement
- Bitumen
- Vinyl ester-styrene
- Soda-lime silica glass.

The radioactive tracers are: Cs137 , Sr85 and Co60. The reference leaching method is a modified ANS 16.1 test (triplicate samples). The leachant, to start with, is distilled water. The specimen size is 4.6 x 6.5 cm. The homogeneity is assured by gamma spectroscopic control checks.

This rather simple system of matrix/tracers/leachant was chosen to produce baseline data, eliminating extraneous variables which may obscure the results. According to the traditional view, the relative order of importance of the mass transport term is as follows (1):

1. Diffusion
2. Dissolution
   Ion exchange   The relative importance may vary
   Corrosion   with the waste type, the
   Surface effects   matrix and th elapsed leach time.
   etc.

Some selected mass transport models are given in Table V.1. To facilitate data processing, a computerized leaching data base is being established pursuing the following steps:

. Determine the "Menu" by which the data are fed into the base
. Provide the retrieval procedure for data processing
. Insert literature and experimental data obtained in this
  programme.

The model will be validated with respect to the chemical and physical
properties, i.e. the leachate analysis (radionuclides, pH, chemical
species, etc.), the waste form analysis before and after leaching, and
the curve fitting using the mass transport models, i.e. the terms in the
mass transport equation must correspond to the process occurring in the
leaching system. This includes optimization of curve fits and statistical analysis.

To predict the long-term behaviour, accelerated leach tests will be
developed. It is necessary to prove that the prevailing mechanisms and
the factors which control the leaching behaviour do not change in
accelerated tests. First, the rate controlling factors will be tested
and evaluated. Some rate controlling factors being considered are:

- temperature
- leachant composition
- leachant pH and Eh
- sample surface to leachant volume ratio
- complexing agents (tannic, humic acid)
- etc.

Then the effects of acceleration factors on the prevailing leaching
mechanisms, with regard to limits and effects of multiple factors, will
be investigated. And finally, the accelerated tests will be validated in
view of:

- Statistical evaluation of the leach results
- Comparison of the leaching behaviour of waste forms with model
  predictions, column experiments and full-scale tests.

The programme spreads well over a 4 to 5 year period. Oak Ridge is
preparing comparative in-situ tests on tracered laboratory and real
waste samples.

To model release mechanisms, Harwell incorporated a zeolitic ion exchan-
ger (clinoptilolite), tracered with Cs137 and Sr90, into OPC. Different
test methods such as: Soxhlet, ISO, low water flow and static tests
were compared over a period of 150 days. The diffusion width of the
isotopes was estimated on the base of a simple diffusion model. The
experimental results proved that the effective diffusion coefficient
is dependent on the intrinsic diffusion characteristics of the material
and the interaction between the radionuclide and the matrix. The model
assumptions constitute a reasonable approach (see Fig. 5.6). The reason
for this approach with a simple model is visualised in Fig. 5.7.
Plotting the fractional release rate as a function of the groundwater
flux shows that the transport of radioactivity is not depending on the
leach rate of radionuclides but rather on their solubility. It also
shows that near field conditions under normal operation can lead to
diffusion controlled release from the repository.

5.2.2 Source-term modelling

5.2.2.1 Cement

In order to set up a model describing the source-term by a finite element method, Fontenay-aux-Roses carries out "integral" experiments under in-situ conditions and varying parameters. With such a model it is anticipated to define the diffusion coefficents for ions, radionuclides at different chemical forms and colloids of interest. Fig. 5.8 gives a schematic view of the approach. The following two waste forms are being tested (simulates of Valduc wastes):

- . Borate concentrates incorporated in Portland cement mortars
- . Nitrates incorporated in Portland cement pastes or mortars.

Different types of cements are also being examined, such as OPC, pozzolanic cements and cements with different additives. The container material is assumed to be concrete. Three materials are considered for the backfill: cement, cement-clay and clay only. For the leachant, different natural waters are being applied. Their composition is given in Table V.2, of which 2 French waters have been tested: the subsoil water of the "Centre de la Manche" and the water from Auriat.
The experimental studies in progress comprise:

- Investigation of corrosion mechanisms
- Measurement of diffusion coefficients.

The corrosive species are Cl^-, SO_4^{2-}, carbonic and humic acids. In order to extrapolate to real disposal conditions, the concentrations have been kept variable, e.g. for Cl^- 0.2 to 20 g/l, for SO_4^{2-} 0.5 to 10 g/l. The initial pH, changing through the various barriers, was 11.5 and 13. Two types of cement, OPC and pozzolanic cement, are currently being tested. The model predictions have been calculated. The diffusion coefficients are being determined in diffusion cells. The ions of concern in a representative pore fluid water (pH > 13) of corroded cement are K^+, Na^+ and Ca^{2+}. The radionuclides which pass through the membrane are being measured under these conditions, as well as in corroded cement as a function of pH and the water chemistry.

Risø carried out diffusion experiments with various types of concrete and found that the diffusion through membranes does not obey Fick's law (see Fig. 3.16). This might be due to enhanced corrosion in the pore structure, to the change during the wetting of the cement membrane and/or to the difference according to which the radionuclides are bound to the high and the low concentration side of the membrane. A possible explanation could be the presence of pore systems penetrating to different depths into the samples.

5.2.2.2 Bitumen

A model describing the leaching mechanisms in bitumen would have to consider, that the amount of water moving in, exceeds the volume of materials released into solution. This leads to swelling which stresses the bitumen film around the waste particles. The thickness of the film decreases with increasing waste loading according to e.g. the diagram given in Fig. 5.9. Stressing the bitumen film, due to water uptake,

leads to defects in the films and consequently to increased ion diffusion and enhances again the water uptake, leading to a multiplying
effect. The swelling stresses can be very high. The behaviour of the
mutual effects, illustrated in Fig. 5.10, should be reflected in the
model. Experiments are being performed in Risø and in other laboratories.

5.2.3 Long-term predictions

For the evaluation of concrete liners for a clay repository, Harwell
pursued a rather unconventional approach. The structural lifetime of a
1 m thick concrete wall (Table V.3) was estimated for different constraints. The major lifetime-limiting factor depends on the choice of
concrete composition, and the transport of dissolved species in the
clay. The estimates indicate that lifetimes of the order of 1000 years
are attainable with a suitable choice of materials. According to these
calculations a low alumina cement-(sulphate resistant) is better than
OPC by a factor of 10. With respect to the interaction of waste and
concrete, only radiation has so far been taken into account.

5.3. Proposals and recommendations (cooperation)

5.3.1 Physico-chemical stability

The physical stability with respect to swelling and/or pressurization
has to be examined more thoroughly under disposal conditions.
This phenomenon is promoted by:

. Radiolysis gases
. Water uptake
. Corrosion or degradation gases (H_2, CO_2).

It could lead to flowing, in case of bitumen, or crumbling, in case of
cement, which affects the specific surface. Water uptake is to be
investigated with respect to its rate, the associated swelling pressure
and the possible engineered countermeasures. In Risø experiments are
also being carried out about the absorption of humidity from air onto
bitumen. Accumulation of leached material on the surface was observed.

In addition, the water transport in multi-barrier systems has to be
studied in detail.
It can be due to:

a) External hydraulic gradients
b) Density differences
c) Osmotic phenomena
d) Concentration differences (diffusion).

For modelling purposes it has to be clarified how to handle:

- Time dependence of protection by steel drums
- Time dependence of pore systems in concrete
- Distribution in pore size and tortuosities
- Effects of gaps between barriers (no backfill, contraction
  defects).

Finally, <u>chemical interactions</u> have to be studied in view of:

i)   Modification of water chemistry depending on the external supply
     of water, ions, CO_2 and C_2 and the amounts of:

     . Leached material from the waste (salts)
     . Leached material from concrete (OH^-, cations)
     . Degradation products (steel, bitumen)

ii) Mutual protection, e.g. steel/concrete, bitumen/concrete

iii) Negative influences or complications due to:

     . K_d (clay, concrete) dependence on water chemistry
     . Speciation dependence on ligand concentration
     . pH and redox conditions
     . Effects of ionic strength on the solubility.

There was a reminder that leachability of tracer isotopes might be
influenced by their chemical form, which could be affected by additives
or impurities, like complexing agents.

5.3.2 Heterogeneous wastes

In many cases heterogeneous wastes are not distributed uniformly in the
matrix. These wastes comprise e.g. decommissioning wastes (moderator
materials, pumps, pipeworks), reprocessing wastes (hulls, graphite),
technological wastes (filter cartridges, Pu-contaminated materials).

At the present time, we do not know the radioactivity release mechanisms
from such waste forms. We can only measure the activity which has
migrated to the leachate, during a leach test. The container, in the
case of heterogeneous wastes takes a leading role in the retention of
radionuclides and must meet established barrier criteria.

Research programmes must be carried out to estimate the source-term
associated with the waste and speciation of solubilized radionuclides to
estimate when a quasi-equilibrium state is established and what the
leach-rate is for this period.

5.3.3 Decontamination effluents

They can contain a large number of different compounds. Synergistic
effects of complexing agents in decontamination wastes on the release
mechanisms are being investigated in Sweden and elsewhere. Their
compatibilty with solidification materials will be examined in the
United States.

5.3.4 Underground facilities

An underground laboratory in clay formation, 240 m below ground, has
been installed on site at Mol. First test assemblies to demonstrate the
compatibility of waste form, barrier material and geological formation
are being introduced early in 1985. The laboratory may become the host
for international collaboration on this topic.

5.3.5 Data base

The BNL data base for release mechnisms, mentioned in § 2.1. page V/3, will provide eventually also lysimetric test results. Tests involving simulated waste forms emplaced in soil columns and leached with site specific groundwater (Barnwell site) under a range of environmental conditions are in progress at BNL. These studies are funded by the U.S. Nuclear Regulatory Commission. The data base will eventually be open for international access.

The EC is preparing a data base for the characterization of radioactive waste forms including leaching characteristics. The structure of the data base has been developed, under the advice of the JRC Ispra and a number of European experts, by the Battelle Institute Frankfurt. Ispra is now trying to ensure that the data in-put and retrievability is in a user - friendly form. During 1985 the experimental data accumulated during the last ten years of coordinated European efforts in this field will be fed in. Subsequently data from the literature and non-European origins, including American data, can be introduced. As soon as the data base is installed the announcement and description will be widely distributed aiming for a large participation of customers.

An interchange between the two data banks, relating to leaching data, is envisaged.

5.4. References

(1) Dougherty, M. Fuhrmann, D.R. and P. Colombo, Leaching Mechanisms Program, BNL 35751, September 1984, Brookhaven National Laboratory, Upton N.Y. 11973.

(2) Dougherty, D.R. and P. Colombo, Leaching Mechanisms of Solidified Low-Level Waste - Literature Survey, BNL 51899, Inpress, Brookhaven National Laboratory, Upton N.Y. 11973.

Table V.1 : Some Selected Mass Transport Models (1)

Model Description		Release Equation for the Mobile Species (a) (Cumulative Fractional Release)
Medium	Transport Mechanism	
semi-infinite, homogeneous, chemically inert	diffusion	$\sum \frac{a_n}{A_o} * \frac{V}{S} = 2\left(\frac{Dt}{\pi}\right)^{1/2}$
semi-infinite, uniform initial concentration	diffusion + concentration-dependent dissolution	$\sum \frac{a_n}{A_o} * \frac{V}{S} = (kD)^{1/2}\left[\left(t + \frac{1}{2k}\right)\operatorname{erf}(kt)^{1/2} + \left(\frac{t}{k\pi}\right)^{1/2} e^{-kt}\right]$
semi-infinite, uniform initial concentration	diffusion + surface dissolution (moving boundary)	$\sum \frac{a_n}{A_o} * \frac{V}{S} = (RD)^{1/2}\left[\left(t + \frac{1}{2R}\right)\cdot\operatorname{erf}(Rt)^{1/2} + \left(\frac{t}{\pi R}\right)^{1/2} e^{-Rt} + t\right]$

a (V/S) = volume to surface ratio; t = time; D = diffusion constant; k = dissolution rate constant; erf = error function; R = $U^2/4D$ where U = velocity on the moving boundary.

Table V.2 : Compositions of Waters Considered
in the Chemical Study

Constituents		Synthetic Water Swedish Standard	Swedish Granitic Underground Water	French Granitic Water (AURIAT)	Water from the La Manche Centre
Na^+	mg/l	65	10 – 100	62 – 99	18 – 80
K^+	"	3,9	1 – 5	2,3 – 7,0	0,6 – 72
Ca^{2+}	"	18	25 – 50	4 – 51	2 – 50
Mg^{2+}	"	4,3	5 – 20	0,4 – 1,7	0,5 – 18
Cl^-	"	70	5 – 50	8,4 – 16,2	
SO_4^{2-}	"	9,6	1 – 15	< 1 – 2,5	
HCO_3^-	"	123	60 – 400	62,5 – 164	
Si_{tot}	"	5,3	1 – 15	3,6	
pH	"	8,0 – 8,2	7,2 – 8,5	9,0 – 9,6	5,5 – 8

Table V.3 : Predicted Structural Lifetime (Years)
for 1 m Thick Concrete

	Shallow Repository (Clay)	Deep Repository (Anhydrite)
Sulphate Attack	380 (OPC) 2,500 (SRPC)*	570 (OPC) 3,800 (SRPC)*
$Ca(OH)_2$ leaching	> 350	>> 350
Reinforcement (Cracking)	900	900
Reinforcement (no cracking)	3,000	3,000

* Sulphate resistant Portland cement

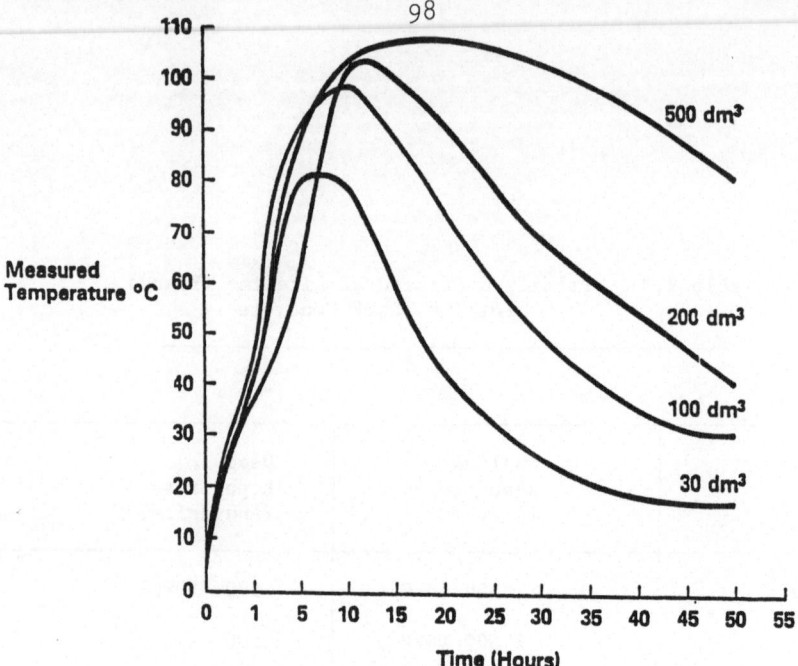

Fig. 5.1 Effect of Container Size on Centre Drum
Temperature during setting of Nitrate/OPC
Mixes

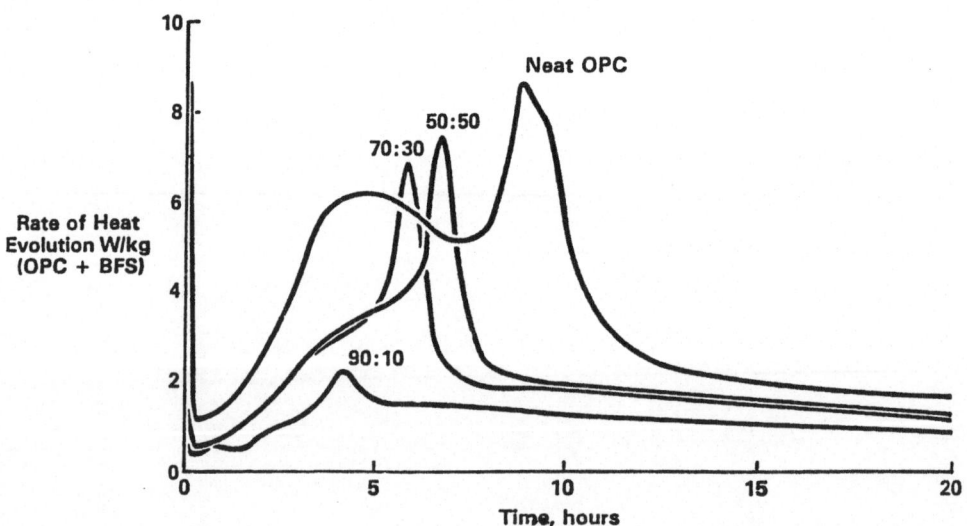

Fig. 5.2 Rate of Heat Evolution for various
BFS : OPC Ratios

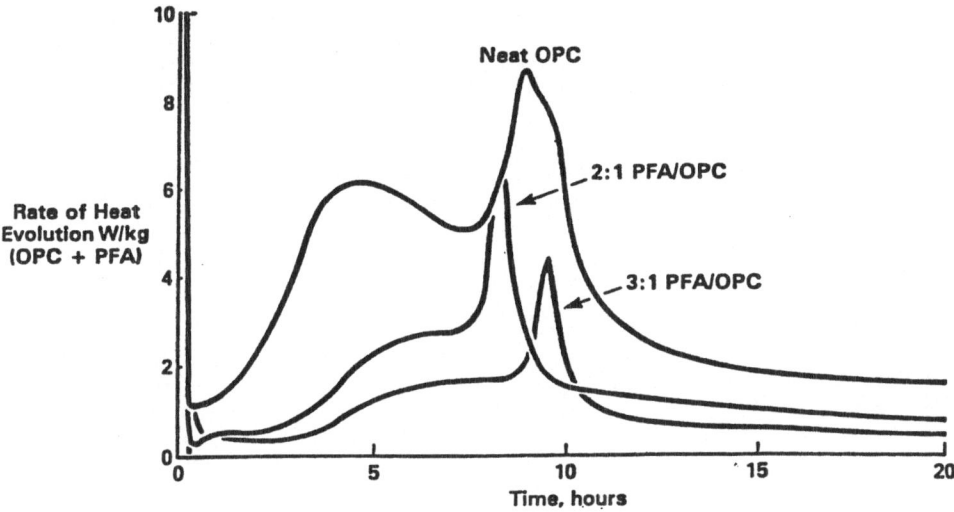

Fig. 5.3 Rate of Heat Evolution for various
PFA : OPC Ratios

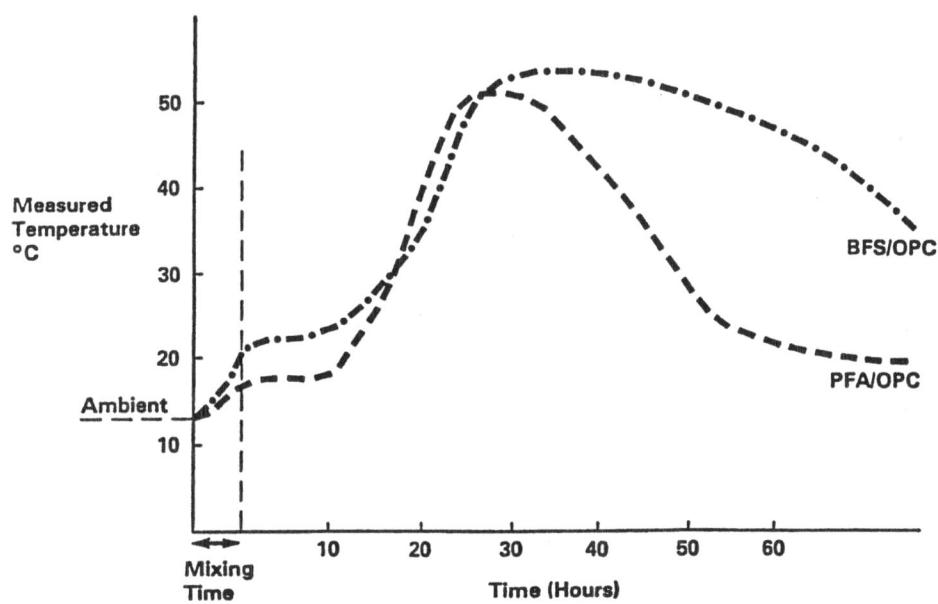

Fig. 5.4 Centre Drum Temperatures during setting of
Cement Grouts on 500 dm³ Scale

100

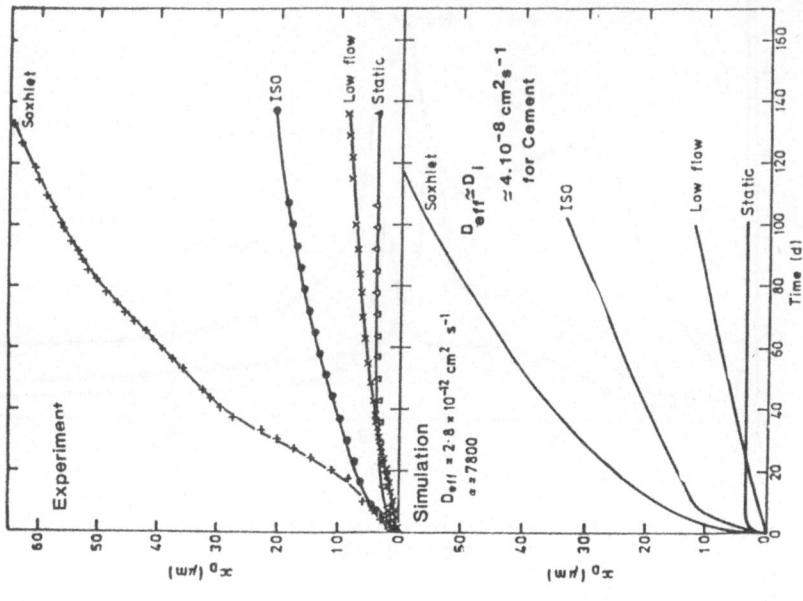

Fig. 5.6 Comparison of Experimental and Predicted Leaching of Cs-137 in OPC/Clinoptilolite, with Different Leach Test Methods

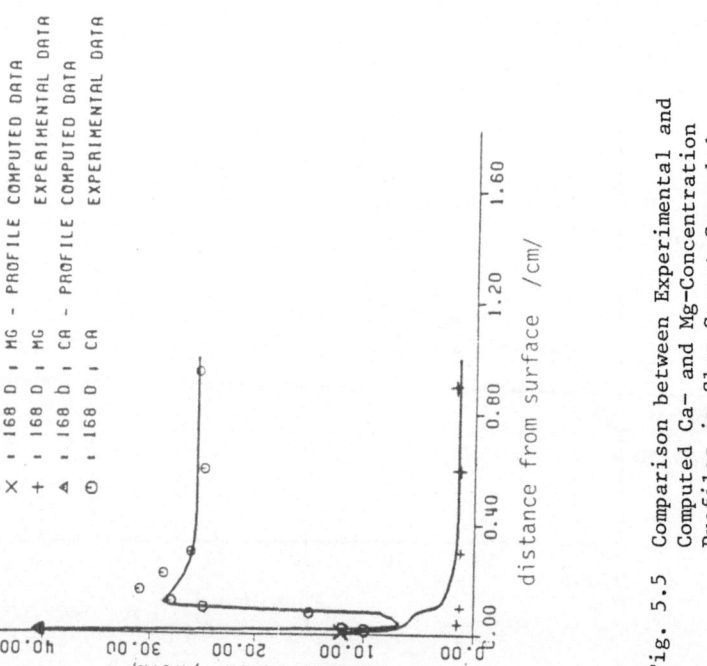

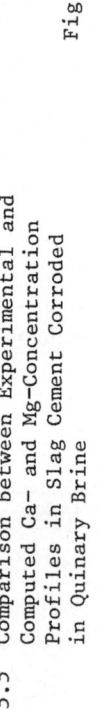

X = 168 D ; MG - PROFILE COMPUTED DATA
+ = 168 D ; MG   EXPERIMENTAL DATA
▲ = 168 D ; CA - PROFILE COMPUTED DATA
⊙ = 168 D ; CA   EXPERIMENTAL DATA

distance from surface /cm/

concentration /wt.%/

Fig. 5.5 Comparison between Experimental and Computed Ca- and Mg-Concentration Profiles in Slag Cement Corroded in Quinary Brine

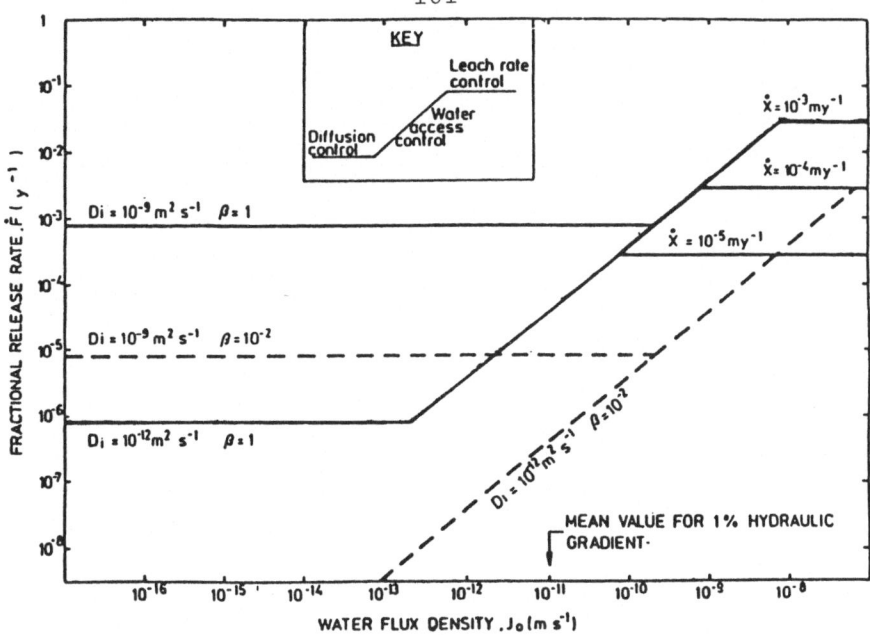

Fig. 5.7 Fractional Release Rate as a function of Groundwater Flux Density of two Nuclides from the Model Repository when Repository and Geology have equal Permeabilities. The Nuclides in their Extent of Solubilization (β)

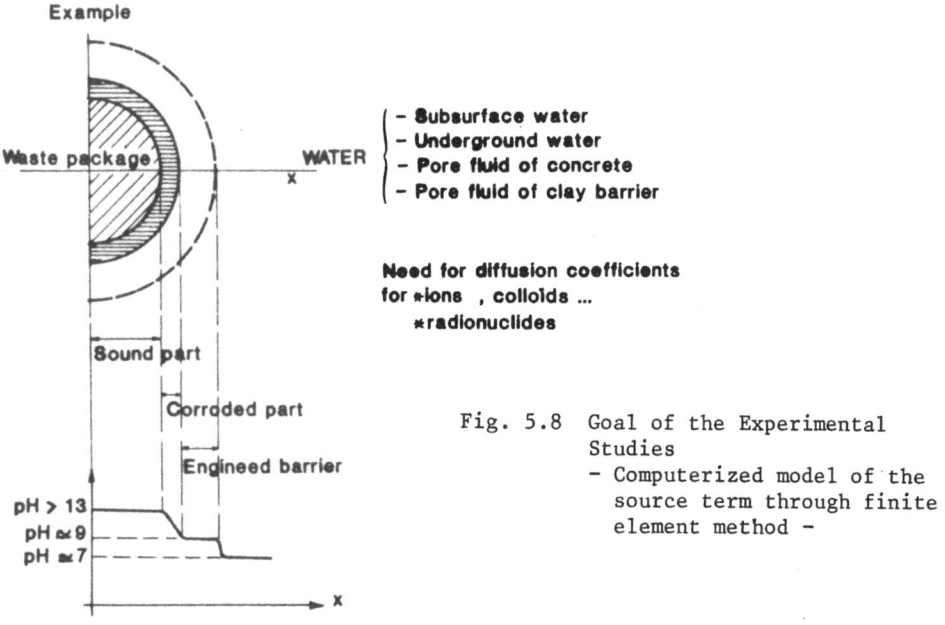

Fig. 5.8 Goal of the Experimental Studies
- Computerized model of the source term through finite element method -

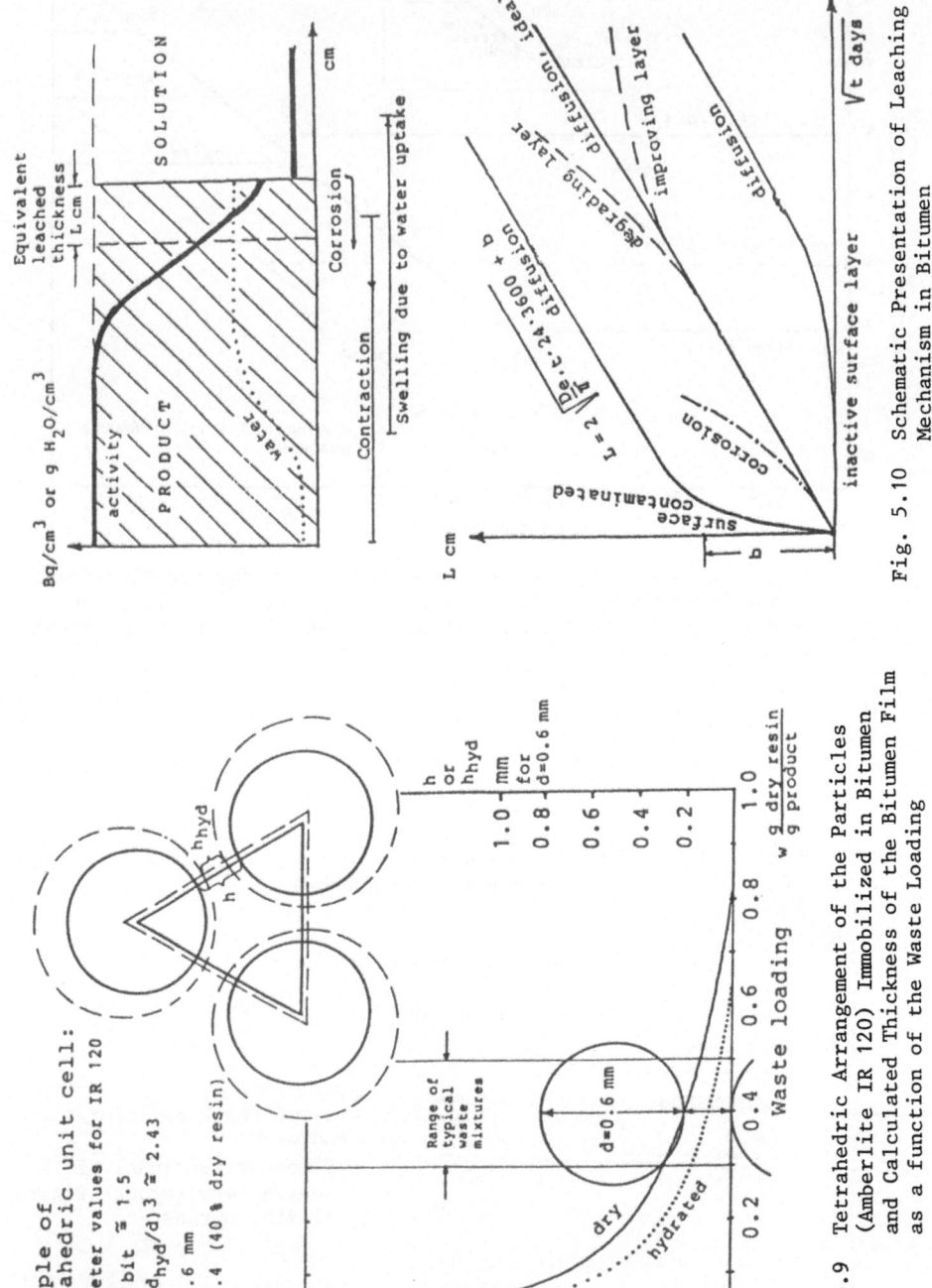

Fig. 5.9 Tetrahedric Arrangement of the Particles
(Amberlite IR 120) Immobilized in Bitumen
and Calculated Thickness of the Bitumen Film
as a function of the Waste Loading

Fig. 5.10 Schematic Presentation of Leaching
Mechanism in Bitumen

CONCLUSION

ACKNOWLEDGEMENTS

LIST OF PARTICIPANTS

6. CONCLUDING REMARKS - FUTURE EVOLUTION

The main objective of this workshop was the evaluation of interest in leach testing results in view of their application as a possible source-term for the risk assessment of LL/MLW disposal sites. Even though several countries believe that the primary concern is the physical behaviour of the waste packages (swelling and gas evolution creating possible accidental conditions), most of the participants agreed that the waste package behaviour in an aqueous environment is the crucial element if risk analysis based on radionuclide migration.

Many informations have been obtained through contributions and discussions giving the present status of existing (or anticipated) regulations and waste form specifications in the different countries (see Session 4). The risk analysis hypotheses have been presented in an extensive "tour de table" mainly during the first Session. According to this, almost all countries are using the multibarrier concept for deep terrestrial disposal. The risk analysis of shallow land burial sites has also been discussed, in particular in relation to possible húman intrusion after site closure.

The different criteria and parameters for waste forms characterization and present data available obtained by various evaluation methods have been discussed during Sessions 2 and 3. Based on the multibarrier concept, the most important effects on the cement, bitumen or polymer waste forms were discussed. The effect of swelling due to irradiation or water uptake has been studied closely and a model has been developed. It has been pointed out that the chemical interactions (including the deleterious effect of chelating agents) must be studied in the entire system: waste/matrix - barriers - host rock. As this is rather difficult to model, a compromise between integral and simplified experiments must be searched for.

Concerning the status of waste form characterization, a wide range of data are available, in particular for homogeneous waste forms. The degree of homogeneity and its signification for experiments has still to be clarified for the so-called "heterogeneous waste forms".

It has been suggested that before starting with accelerating tests (yet by undefined methods) preference should be given to study, with increased emphasis, the release mechanisms, e.g. the microstructural changes due to ageing in order to improve the basis for modelling. Recommendations for future programmes have been presented in Session 5. It has been emphasized that international cooperation in the following topics is required: physical stability, water transport and chemical interactions. A data base, like the one proposed by the CEC, would be of great help for this purpose.

Unique in its nature, this workshop rendered it possible to the participants (who were all experts in their propre countries) to have three full days of common discussion. The proceedings proved that present results of the aqueous behaviour of LL/MLW are significant and that certain aspects (lysimetry, waste/barrier interaction, modelling..) have to be studied more closely. It also was pointed out that contacts between the following three main areas of radwaste management should be reinforced: waste characterization, risk analysis and waste disposal. This kind of workshop could be useful to be held in future under the same conditions (restricted attendance, no presubmitted papers) and at a frequency of about two years. Brookhaven National Laboratory (Upton, NY-USA) has recently agreed to host the 2nd Seminar, in the frame of the US-DOE/CEC cooperation, in fall 1986.

Thanking for the collaboration of participants, we wish a full success of the second workshop.

7. ACKNOWLEDGEMENTS

The organising members would like to express their thanks to Division XII of the CEC and the DERDCA of CEA for the preparation of the workshop and the publication of the proceedings.

We also like to thank the different contributors who accepted to revise the draft proceedings and to submit their comments. In particular we like to thank R.A.J. Sambell for the revising of the entire draft and H. Forsström (KBS-Sweden) and E. Warnecke (PTB-FRG), who could not attend the seminar but forwarded nevertheless their contribution.

We also thank the two scientific secretaries for their almost one year effort in preparing the workshop and the proceedings.

LIST OF PARTICIPANTS

ALDER J.C.	NAGRA/CEDRA Parkstrasse 23 CH - 5401 BADEN SWITZERLAND
ATABEK R.M.(Mrs)	CEA/Centre d'Etudes Nucléaires de Fontenay-aux-Roses. Département de Recherche et Développement Déchets/SESD. B.P. N° 6 - 92260 Fontenay-aux-Roses FRANCE
ATKINSON A.	UKAEA/HARWELL AERE HARWELL, OXFORDSHIRE OX11 ORA UNITED KINGDOM
BARRETT L.	US. NRC 7915 EASTERN Avenue SILVER SPRING 20555 WASHINGTON DC UNITED STATES OF AMERICA
BAUDIN G.	CEA/Centre d'Etudes Nucléaires de Fontenay-aux-Roses. Département de recherche et Développement Déchets. B.P. N° 6 - 92260 Fontenay-aux-Roses FRANCE
BOURDREZ A.	CEA/ANDRA 29-33 rue de la Fédération 75015 PARIS FRANCE

BRODERSEN K.

Risø National Laboratory
Postbox 49
DK - 4000 ROSKILDE
DENMARK

CADELLI N.

Commission of European Communities
DG XII/DI
200, rue de la Loi
1049 BRUXELLES
BELGIUM

CHAPUIS A.M. (Mrs)

CEA/Centre d'Etudes Nucléaires de Fontenay-
aux-Roses. Institut de Protection et de sûreté
Nucléaire/DPT.
B.P. N° 6 - 92260 Fontenay-aux-Roses
FRANCE

CLAES J.

ONDRAF/NIRAS
Bd du Régent 54 Boite 5
1000 BRUXELLES
BELGIUM

COLOMBO P.

Department of Nuclear Energy - B.N.L.
Upton, Long Island
NEW-YORK 11973
UNITED STATES OF AMERICA

DAYAL R.

Nuclear Waste Research Group - B.N.L.
Upton, Long Island
NEW-YORK 11973
UNITED STATES OF AMERICA

DE BATIST R.

SCK/CEN MOL
BOERETANG 200
B - 2400 MOL
BELGIUM

DONATO A.

Energia Nucleare ed Energie Alternative
CRE Casaccia, S.P. Anguillarese km 1300
I - 00060 S. Maria di Galeria - ROMA
ITALY

GILLIAM T.M.

ORNL
P.O. BOX X - OAK RIDGE
TENNESSEE 37831
UNITED STATES OF AMERICA

IWAMOTO F.

Nuclear Research Center J.G.C.
2205 Naritacho Oharaimachi Higashi-Ibaraki-Gun
IBARAKI Pref - 311-13
JAPAN

JOURDE P.

CEA/Centre d'Etudes Nucléaires de Fontenay-
aux-Roses. Direction des Effluents et Déchets.
B.P. N° 6 - 92260 Fontenay-aux-Roses
FRANCE

KOSTER R.

Kernforschungszentrum Karlsruhe GmbH
Postfach 3640
7500 KARLSRUHE
FEDERAL REPUBLIC OF GERMANY

LEWI J.

CEA/Centre d'Etudes Nucléaires de Fontenay-
aux-Roses. Institut de Protection et de Sûreté
Nucléaire/DAS - B.P. N° 6
92260 Fontenay-aux-Roses
FRANCE

MENDEL J.E.

BATTELLE - PNL
Battelle boulevard - Richland
WASHINGTON 99352
UNITED STATES OF AMERICA

NOMINE J.C.

CEA/Centre d'Etudes Nucléaires de Saclay.
Département de Recherche et Développement
Déchets/SESD. B.P. N° 2
91120 GIF-SUR-YVETTE
FRANCE

ODOJ R.

K.F.A. Jülich GmbH
Postfach 1913
5170 JÜLICH
FEDERAL REPUBLIC OF GERMANY

POTTIER P.

CEA/Centre d'Etudes Nucléaires de Cadarache
Département de Recherche et Développement
Déchets.
13108 Saint-Paul-lez-Durance Cedex
FRANCE

PRICE MTS.

UKAEA/AERE Winfrith
Technology Branch Bdg A 32 - DORCHESTER
DORSET, DT 28 DH
UNITED KINGDOM

SAAS A.

CEA/Centre d'Etudes Nucléaires de Cadarache
Département de Recherche et Développement
Déchets/BECC.
13108 Saint-Paul-lez-Durance Cedex
FRANCE

SAMBELL R.A.J.

Materials Development Division
AERE HARWELL, OXFORDSHIRE
OX 11 ORA
UNITED KINGDOM

SIMON R.

Commission of The European Communities
DG XII/DI - 200, rue de la Loi
1049 BRUXELLES
BELGIUM

SJOEBLOM R.

STUDSVIK Energiteknik AB
S 611 82 NYKÖPING
SWEDEN

STEARN S.

Department of the Environment
Room A5.33
Romney House, 43 Marsham Street
LONDON SW 1 P 3 PY
UNITED KINGDOM

WILLIAMSON A.S.

ONTARIO HYDRO
800 Kipling Avenue
TORONTO M8 Z 554 - ONTARIO
CANADA

Scientific Secretariat :

DOZOL M.(Mme)

CEA/Centre d'Etudes Nucléaires de Cadarache.
Département de Recherche et Développement
Déchets/CIDN
13108 Saint-Paul-lez-Durance Cedex
FRANCE

KRISCHER W.

Commission of the European Communities
DG XII/DI - 200, rue de la Loi
1049 BRUXELLES
BELGIUM